Edited By
Carl Goodson
Barbara Adams
Marsha Sneed

Fire Service Ventilation

Seventh Edition

Validated By
THE INTERNATIONAL FIRE SERVICE TRAINING ASSOCIATION

Published By
FIRE PROTECTION PUBLICATIONS
Oklahoma State University

ISBN 0-87939-109-X
Library of Congress 94-70300

Seventh Edition
First Printing, March 1994
Second Printing, October 1996
Third Printing, August 2000

Printed in the United States of America

Dedication

This manual is dedicated to the members of that unselfish organization of men and women who hold devotion to duty above personal risk, who count on sincerity of service above personal comfort and convenience, who strive unceasingly to find better ways of protecting the lives, homes and property of their fellow citizens from the ravages of fire and other disasters . . . ***The Firefighters of All Nations.***

Dear Firefighter:

The International Fire Service Training Association (IFSTA) is an organization that exists for the purpose of serving firefighters' training needs.Fire Protection Publications is the publisher of IFSTA materials. Fire Protection Publications staff members participate in the National Fire Protection Association and the International Association of Fire Chiefs.

If you need additional information concerning our organization or assistance with manual orders, contact:

Customer Services
Fire Protection Publications
Oklahoma State University
930 N. Willis
Stillwater, OK 74078-8045
1 (800) 654-4055

For assistance with training materials, recommended material for inclusion in a manual, or questions on manual content, contact:

Technical Services
Fire Protection Publications
Oklahoma State University
930 N. Willis
Stillwater, OK 74078-8045
(405) 744-5723

The International Fire Service Training Association (IFSTA) was established as a "nonprofit educational association of fire fighting personnel who are dedicated to upgrading fire fighting techniques and safety through training." This training association was formed in November 1934, when the Western Actuarial Bureau sponsored a conference in Kansas City, Missouri. The meeting was held to determine how all the agencies interested in publishing fire service training material could coordinate their efforts. Four states were represented at this initial conference. Because the representatives from Oklahoma had done some pioneering in fire training manual development, it was decided that other interested states should join forces with them. This merger made it possible to develop training materials broader in scope than those published by individual agencies. This merger further made possible a reduction in publication costs, because it enabled each state or agency to benefit from the economy of relatively large printing orders. These savings would not be possible if each individual state or department developed and published its own training material.

To carry out the mission of IFSTA, Fire Protection Publications was established as an entity of Oklahoma State University. Fire Protection Publications' primary function is to publish and disseminate training texts as proposed and validated by IFSTA. As a secondary function, Fire Protection Publications researches, acquires, produces, and markets high-quality learning and teaching aids as consistent with IFSTA's mission. The IFSTA Executive Director is officed at Fire Protection Publications.

IFSTA's purpose is to validate training materials for publication, develop training materials for publication, check proposed rough drafts for errors, add new techniques and developments, and delete obsolete and outmoded methods. This work is carried out at the annual Validation Conference.

The IFSTA Validation Conference is held the second full week in July, at Oklahoma State University or in the vicinity. Fire Protection Publications, the IFSTA publisher, establishes the revision schedule for manuals and introduces new manuscripts. Manual committee members are selected for technical input by Fire Protection Publications and the IFSTA Executive Secretary. Committees meet and work at the conference addressing the current standards of the National Fire Protection Association and other standard-making groups as applicable.

Most of the committee members are affiliated with other international fire protection organizations. The Validation Conference brings together individuals from several related and allied fields, such as:

- Key fire department executives and training officers
- Educators from colleges and universities
- Representatives from governmental agencies
- Delegates of firefighter associations and industrial organizations
- Engineers from the fire insurance industry

Committee members are not paid nor are they reimbursed for their expenses by IFSTA or Fire Protection Publications. They come because of commitment to the fire service and its future through training. Being on a committee is prestigious in the fire service community, and committee members are acknowledged leaders in their fields. This unique feature provides a close relationship between the International Fire Service Training Association and other fire protection agencies, which helps to correlate the efforts of all concerned.

IFSTA manuals are now the official teaching texts of most of the states and provinces of North America. Additionally, numerous U.S. and Canadian government agencies as well as other English-speaking countries have officially accepted the IFSTA manuals.

Table Of Contents

Preface

This seventh edition of **Fire Service Ventilation** was written not only to be consistent with the third edition of **Essentials of Fire Fighting** but also to provide information on the theory and practice of fire service ventilation in greater depth than that provided in **Essentials**. This manual addresses those objectives in NFPA 1001, *Standard on Fire Fighter Professional Qualifications (1992)*, that apply to ventilation.

Acknowledgement and special thanks are extended to the members of the IFSTA validating committee who contributed their time, wisdom, and knowledge to this manual.

Chairman
Paul H. Boecker
Lisle-Woodridge Fire District
Lisle, IL

Secretary
Robert Anderson
Spokane County Fire District #9
Spokane, WA

David F. Clark
Illinois Fire Service Institute
Champaign, IL

Dexter Coffman
Tempest Technology Corporation
Fresno, CA

Tom Hebert
LSU Firemen Training Program
Baton Rouge, LA

John Hoglund
Maryland Fire & Rescue Institute
College Park, MD

John Norman
Nassau County Fire Service Academy
Inwood, NY

Others who generously contributed their time and expertise to the committee were:

Don Heiser
Encinitas Fire Department
Encinitas, CA

Clyde Mariotti
Tri-State Fire Protection District
Darien, IL

Special recognition is given to the Seattle (WA) Fire Department whose help and cooperation were essential in obtaining a large number of the photographs needed to illustrate this manual. Helping with the project were:

Fire Chief Claude Harris
Deputy Chief Stewart Rose
Lieutenant Jesse Atterberry

Also instrumental in helping us obtain photos were the Bainbridge Island (WA) Fire District #2, under the direction of Acting Director Dean Thoemke, and the Yukon (OK) Fire Department, under the direction of Chief Robert Noll.

The following individuals and organizations contributed information, photographs, or other assistance that made the completion of this manual possible:

Bilco Company, New Haven, CT
Cutters Edge, Julian, CA
Harvey Eisner, Tenafly (NJ) Fire Department
Ray C. Goad, Goad Engineering, Boonville, IN
Ron Jeffers, New Jersey Metro Fire Photographers Association
Wes Kitchel, Santa Rosa (CA) Fire Department
Lisle-Woodridge Fire District (IL)
City of Los Angeles (CA) Fire Department
Joseph J. Marino, New Britain, CT
Tom McCarthy, Chicago (IL) Fire Department
Andy Maxey, Oklahoma State University
Bob Norman, Singerly VFD, Elkton (MD)
Oberlin World Color Press, Stillwater, OK
Edward Prendergast, Chicago (IL) Fire Department
RAMFAN Corporation, National City, CA
Bob Ramirez, Los Angeles (CA) Fire Department (Retired)
Reliable Fire Equipment Company
Douglas Shelby
Wayne Sibley, Dallas/Ft. Worth International Airport
Super Vac Corpration
Survivair, Santa Ana, CA
Tempest Corporation
Jim Vencil, Oklahoma State University
Barry Wagner, Illinois Fire Service Institute
Joel Woods, Maryland Fire & Rescue Institute

Finally, gratitude is also extended to the following members of the Fire Protection Publications staff whose contributions made the final publication of this manual possible:

Michael Wieder, Senior Publications Editor
Carol Smith, Senior Publications Editor
Cynthia Brakhage, Associate Editor
Susan S. Walker, Coordinator of Instructional Development
John Hoss, Associate Editor
Don Davis, Production Coordinator
Shari Downs, Graphic Designer
Ann Moffat, Senior Graphic Designer
Lori Williamson, Graphic Designer
Desa Porter, Graphic Designer
Kimberly Edwards, Photographic Technician
Pat Agostinelli, Research Technician
Ward Barnett, Research Technician
Jason Goetz, Research Technician
Matthew Manfredi, Research Technician
Mark Riggbee, Research Technician

Lynne C. Murnane

Lynne C. Murnane, Managing Editor

Glossary

A

Aerial Apparatus
Any piece of mobile fire apparatus having the means to reach at least 65 feet (20 m) above the ground, with many being capable of reaching more than 100 feet (30 m). Aerial devices are usually either aerial ladders or telescoping or articulating booms.

Air-Handling System
See HVAC.

Arched Roof
Any of several different types of roofs of which all are curved or arch shaped, resembling the top half of a horizontal cylinder. Typical applications are on supermarkets, auditoriums, bowling centers, sports arenas, and aircraft hangars.

Area Of Refuge
Two-hour-rated compartment containing one elevator to the ground floor and at least one enclosed exit stairway.

Assembly
All component parts necessary to form a complete unit or system joined or fitted together.

Atrium
Open area in the center of a building, similar to a courtyard but usually covered by a skylight, to allow natural light and ventilation into interior rooms.

Attic
Concealed and often unfinished space between the ceiling of the top floor and the roof of a building. Also called Cockloft or Interstitial Space.

Autoexposure
See Lapping.

Awning Window
Window that is hinged at the top and swings outward at the bottom.

B

Backdraft
Instantaneous explosion or rapid burning of superheated gases in a confined space. It may occur because of inadequate or improper ventilation procedures.

Balloon-Frame Construction
Type of framing used in some single-story and multistory buildings wherein the studs are continuous from the foundation to the roof. There may be no fire-stops between the studs.

Beam
Structural member in a roof or floor assembly that is subjected to loads perpendicular to its length.

Bearing Wall
Wall that supports itself and the weight of the roof and/or other structural components above it.

Billy Pugh Net
Rope net or basket, designed to be suspended beneath a helicopter, for transporting personnel and/or equipment.

Blower
Large-volume fan used to blow fresh air into a building or other confined space. Often used in positive-pressure ventilation (PPV), blowers are most often powered by gasoline engines, but some have electric motors.

Breach
To make an opening (usually in a masonry wall) to allow access or for ventilation.

Brick Veneer
Single layer of bricks applied to the inside or outside surface of a wall for aesthetic and/or insulation purposes.

Bridge Truss
Heavy-duty truss, usually made of heavy wooden members with steel tie rods, that has horizontal top and bottom chords and steeply sloped ends.

Building Engineer
That person who is familiar with and responsible for the operation of a building's heating, ventilating, and air-conditioning (HVAC) system and other essential equipment.

Built-Up Roof
Roof covering made of several alternate layers of roofing paper and tar, with the final layer of tar being covered with pea gravel or crushed slag. Also called Tar And Gravel Roof.

Bulkhead
Structure on the roof of a building through which the interior stairway opens onto the roof.

Butterfly Roof
V-shaped roof style resembling two opposing shed roofs joined along their lower edges.

C

Casement Window
Window hinged along one side, usually designed to swing outward by means of a small crank, with the screen on the inside.

Ceiling
Nonbearing structural component separating a living/working space from the underside of the floor or roof immediately above.

Cement
Any adhesive material. In construction, a clay mixture called portland cement is combined with sand and/or other aggregates and water to produce mortar or concrete.

Center Rafter Cut
See Louver Vent.

Chase
See Pipe Chase.

Chimney Effect
See Stack Effect.

Chord
Top or bottom longitudinal member of a truss.

Churning
Movement of smoke being blown out of a ventilation opening only to be drawn back inside by the negative pressure created by the ejector because the open area around the ejector has not been sealed.

Cockloft
See Attic.

Compression
Those vertical or horizontal forces that tend to press things together; for example, the force exerted on the top chord of a truss.

Concealed Space
Any structural void that is not readily visible from a living/working space within a building.

Concrete
Mixture of portland cement and an aggregate filler/binder to which water is added to form a rigid building material. In structural concrete the filler/binder is usually sand and/or gravel; in lightweight concrete, such as soundproofing material, the filler/binder may be sand and/or vermiculite.

Conduction
Transfer of heat by direct contact or through an intervening heat-conducting medium.

Convection
Transfer of heat by the movement of fluids or gases, usually in an upward direction.

Crawl Space
Space between the ground and the floor, a space between the ceiling and the roof, or any other structural void with a vertical dimension that does not allow a person to stand erect within the space. These spaces often contain ductwork, plumbing, wiring, etc.

Curtain Boards
Fire-resistive half-walls, intended to limit the spread of smoke and fire gases, that extend down from the underside of the roof of some commercial buildings. Also called Draft Stop.

D

Dead Load
Weight of the structure, structural members, building components, and any other feature that is constant and immobile.

Decking
See Sheathing.

Dome Roof
Hemispherical roof assembly, usually supported only at the outer walls of a circular or many-sided structure.

Double-Hung Window
Window having two vertically moving sashes.

Draft Stop
See Curtain Boards.

Drywall
See Wallboard.

E

Elevating Platform
Work platform attached to the end of an articulating or telescoping aerial device.

Entry Point (Entry Opening)
In ventilation, the opening through which replacement air enters; usually the same opening through which rescue/attack crews enter the structure.

Exit Opening
In ventilation, the opening that is made or used to release heat, smoke, and other contaminants to the atmosphere.

Exterior Exposure
Building or other combustible object located close to the fire building that is in danger of becoming involved due to heat transfer from the fire building.

F

False Roof
Fascia or facade added to some buildings with flat roofs to create the appearance of a mansard roof.

Fan
Generic term used interchangeably for both blowers and smoke ejectors.

Fiberboard
Lightweight insulation board made of compressed cellulose fibers and often used in suspended ceilings.

Fire Door
Rated assembly designed to automatically close and cover a doorway in a fire wall during a fire.

Fire Escape
Means of escaping from a building in case of fire; usually an interior or exterior stairway or slide independently supported and made of fire-resistive materials.

Fire-Stop
Solid material, such as a wood block, placed within a wall void to retard or prevent the spread of fire through the void.

Fire Wall
Rated separation wall, usually extending from the foundation up to and through the roof of a building, to limit the spread of a fire.

Fixed Window
Window not designed to be opened.

Flashover
Stage of a fire at which all surfaces and objects within a space have been heated to their ignition temperature and flame breaks out almost at once over the surface of all objects in the space.

Flat Roof
Roof with little or no pitch or slope. Flat roofs generally have a slight pitch to facilitate runoff.

Forced Ventilation
Any means of ventilation other than natural. This type of ventilation may involve the use of blowers, smoke ejectors, or fog streams. Also called Mechanical Ventilation.

G

Gabled Roof
Style of pitched roof with square ends in which the end walls of the building form triangular areas beneath the roof.

Gambrel Roof
Style of gabled roof on which each side slopes at two different angles; often used on barns and similar structures.

Gang Nail
Form of gusset plate. These thin steel plates are punched with acutely V-shaped holes that form sharp prongs on one side that penetrate wooden members to fasten them together.

Girder
(1) Large, horizontal structural member used to support joists and beams at isolated points along their length. (2) Steel frame members in a bus that run from front to rear to strengthen and shape the roof bows.

Grade
Natural, unaltered ground level.

Gusset Plate
Metal or wooden plates used to bind roof or floor components into a load-bearing unit.

H

Hatch
Refers to a square or rectangular access opening on a roof, usually locked on the inside. Also see Scuttle.

High Rise
Uniform Building Code (UBC) definition: any building of more than 75 feet (23 m) in height. Practical definition: any building with a roof that is beyond the reach of the fire department's tallest ladder.

Hip Roof
Pitched roof that has no gables. All facets of the roof slope down from the peak to an outside wall.

Hopper Window
Window hinged along the bottom edge and usually designed to open inward.

Horizontal Ventilation
Any technique by which heat, smoke, and other products of combustion are channeled horizontally out of a structure by way of existing or created openings.

HVAC
Heating, ventilating, and air-conditioning system within a building and the fans, ducts, dampers, and other equipment necessary to make it function. Usually a single, integrated unit with a complex system of ducts throughout the building. Also called Air-Handling System.

I

I Beam
Steel or wooden structural member with a cross section resembling a capital I.

Incident Commander (IC)
Person in overall charge of an emergency.

Index Gas
Any commonly encountered gas, such as carbon monoxide (CO) in fires, whose concentration can be measured. In the absence of devices capable of measuring the concentrations of other gases present, the CO measurement may be assumed to indicate their concentrations as well.

Interior Exposure
Areas of a fire building that are not involved in fire but are connected to the fire area in such a manner that may facilitate fire spread through any available opening.

Interstitial Space
See Attic.

Inversion
Atmospheric phenomenon that allows smoke to rise until its temperature equals the surrounding air temperature, and then it spreads laterally in a horizontal layer. Also called Night Inversion.

J

Jalousie Window
Window consisting of narrow, frameless glass panes set in metal brackets at each end that allow the panes a limited amount of axial rotation for ventilation.

Joist
Horizontal supporting member in a roof, ceiling, or floor assembly.

L

Labeled Assembly
See Rated Assembly.

Lantern Roof
Roof style consisting of a high, gabled roof with a vertical wall above a downward-pitched shed roof section on either side.

Lapping
Means by which fire spreads vertically from floor to floor in a multistory building. Fire issuing from a window laps up the outside of the building and enters the floor(s) above, usually through the windows. Also called Autoexposure.

Lee (Leeward)
Direction opposite that from which the wind is blowing; the protected side.

Light Well (Light Shaft)
Vertical shaft at or near the center of a building to provide natural light and/or ventilation to offices or apartments not located on an outside wall.

Live Load
Loads within a building that are movable. Merchandise, stock, furnishings, occupants, firefighters, and the water used for fire suppression are examples of live loads.

Lobby Control
In high-rise fire fighting, the person responsible for taking and maintaining control of the lobby and the elevators, setting up internal communications, coordinating the flow of personnel and equipment up the interior stairway(s), and coordinating with the building engineer.

Louver Vent
Rectangular exit opening cut in a roof, allowing a section of roof deck (still nailed to a center rafter) to be tilted, thus creating an opening similar to a louver. Also called Center Rafter Cut.

Low-Rise Elevator
Elevator that does not serve the fire floor or above in any multistory building. These are often elevators that serve only the lower floors of a high-rise building.

M

Mansard Roof
Roof style with characteristics similar to both gambrel and hip roofs. Mansard roofs have slopes of two different angles, and all sides slope down to an outside wall.

Mechanical Ventilation
See Forced Ventilation.

Membrane Roof
Roof covering consisting of a single layer of waterproof synthetic membrane over one or more layers of insulation on a roof deck. Also called Single-Ply Roof.

Modern Mansard Roof
Roof style having sides that slope at only one angle up to meet a flat deck in the center section.

Monitor Appliance (Monitor)
Master stream appliance whose stream direction can be changed while water is being discharged. Monitors may be fixed, portable, or a combination.

Monitor Roof
Roof style similar to an exaggerated lantern roof having a raised section along the ridge line, providing additional natural light and ventilation.

Monitor Vent
Structure, usually rectangular in shape, that penetrates the highest point of a roof to provide additional natural light and/or ventilation. They may have metal, glass, wired glass, or louvered sides.

Mushrooming
Tendency of heat, smoke, and other products of combustion to rise until they encounter a horizontal obstruction. At this point they will spread laterally until they encounter vertical obstructions and will begin to bank downward.

N

Natural Ventilation
Ventilation techniques that use the wind, convection currents, and other natural phenomena to ventilate a structure without using fans, blowers, or other mechanical devices.

Negative-Pressure Ventilation (NPV)
Technique using smoke ejectors to develop artificial circulation and to pull smoke out of a structure. Smoke ejectors are placed in windows, doors, or

roof vent holes to pull the smoke, heat, and gases from inside the building and eject them to the exterior.

Neutral Pressure Plane

That point within a building, especially a high rise, where the interior pressure equals the atmospheric pressure outside. This plane will move up or down, depending on variables of temperature and wind.

Night Inversion

See Inversion.

Nonbearing Wall

Wall, usually interior, that supports only its own weight. These walls can be breached or removed without compromising the structural integrity of the building.

O

Occupancy

Uniform Building Code (UBC) classification of the use to which owners or tenants put buildings or portions of buildings.

Open Up

To ventilate a building or other confined space.

P

Parapet Wall

Extension of the exterior walls above the roof; also, any required fire walls surrounding or dividing a roof or surrounding roof openings such as light or ventilation shafts.

Party Wall

Wall common to two buildings.

Penthouse

Enclosure on the roof of a building, but not extending more than 12 feet (4 m) above the roof, used primarily for living or recreational purposes.

Pipe Chase

Concealed vertical channel in which pipes and other utility conduits are housed. Pipe chases that are not properly protected can be major contributors to the vertical spread of smoke and fire in a building. Also called Chase.

Pitched Roof

Roof, other than a flat or arched roof, having one or more pitched or sloping surfaces.

Plasterboard

See Wallboard.

Plate

Horizontal construction member at the bottom (soleplate) or top (top plate) of a framed wall.

Platform Frame Construction

Type of framing in which each floor is built as a separate platform, and the studs are not continuous from floor to floor. Also called Western Frame Construction.

Poke-Through

Opening in a floor, ceiling, or wall through which ductwork, plumbing, or electrical conduits pass. If these openings are not properly caulked or sealed, they can contribute significantly to the spread of smoke and fire in a building.

Positive-Pressure Ventilation (PPV)

Method of ventilating a confined space by mechanically blowing fresh air into the space in sufficient volume to create a slight positive pressure within, thereby forcing the contaminated atmosphere out the exit opening.

Pre-Incident Inspection

Thorough and systematic inspection of a building for the purpose of identifying significant structural and/or occupancy characteristics to assist in the development of a pre-incident plan.

Pre-Incident Plan

Operational plan for the safe and efficient handling of emergency situations within a specific building or occupancy.

Pressure Differential

Effect of altering the atmospheric pressure within a confined space by mechanical means. When air is exhausted from within the space, a low-pressure environment is created, and replacement air will be drawn in. When air is blown into the space, a high-pressure environment is created, and air within will move to the outside.

R

Radiation
Transfer of heat through intervening space by infrared thermal waves.

Rafter
Usually sloping wooden roof member to which the roof decking is nailed.

Rated Assembly
Refers to doors, walls, roofs, and other structural features that may be, because of the occupancy, required by code to have a minimum fire-resistance rating from an independent testing agency. Also called Labeled Assembly.

Razor Ribbon
Coil of lightweight, flexible metallic ribbon with extremely sharp edges; often installed on parapet walls and on fence tops to discourage trespassers.

Reading A Roof
Process of observing important features of a roof from a point of safety in order to assess the roof's condition before stepping onto it.

Refuse Chute
Vertical shaft with a self-closing access door on every floor, usually extending from the basement or ground floor to the top floor in multistory buildings.

Rehab
Incident Command System (ICS) term for a rehabilitation station at a fire or other incident where personnel can rest, rehydrate, and recover from the stresses of the incident.

Ridge Board
Highest horizontal member in a pitched roof to which the upper ends of the rafters attach. Also called Ridge Beam or Ridgepole.

Rollover
Condition that may develop in the early steady-state phase of a fire in a confined space. When additional oxygen is supplied by opening doors and/or applying fog streams, superheated unburned gases that have accumulated at the ceiling ignite and roll across the ceiling of the space.

Roof Decking
See Sheathing.

S

Sawtooth Roof
Roof style characterized by a series of alternating vertical walls and sloping roofs, resembling the teeth of a saw, most often found on industrial buildings.

Scissor Stairs
Two sets of crisscrossing stairs in a common shaft, with each set serving every floor but on alternately opposite sides of the stairshaft. For example, one set would serve the west wing on even-numbered floors and the east wing on odd-numbered floors, while the other set would serve floors opposite to the first set.

Scuttle
Openings in ceilings or roofs, fitted with removable covers, that may be used for access or ventilation.

Shaft
Any vertical enclosure within a building; for example, a stairwell, an elevator hoistway, etc.

Shear Strength
Ability of a building component or assembly to resist lateral or shear forces.

Sheathing
Refers to the first layer of roofing material laid directly over the rafters or other roof supports. Sheathing may be plywood or chipboard sheets or may be planks that are butted together or spaced about 1 inch (25 mm) apart. Also called Decking or Roof Decking.

Shed Roof
Pitched roof with a single, sloping aspect, resembling half of a gabled roof.

Sheetrock® Plaster Wallboard
See Wallboard.

Single-Ply Roof
See Membrane Roof.

Size-Up
Ongoing evaluation of an emergency situation done mentally by the officer in charge, resulting in a plan of action that may be adjusted as the situation changes.

Skylight
Any of a variety of roof structures or devices intended to increase natural illumination within buildings.

Smoke Ejector
Ducted fan used primarily to exhaust smoke from a building, but it may be used to blow in fresh air. Most are driven by electric motors, but some are driven by gasoline engines or by water pressure. They may be used in conjunction with a flexible duct.

Smoke Tower
Fully enclosed escape stairway that exits directly onto a public way. These enclosures are either mechanically pressurized or require the user to exit the building onto an outside balcony before entering the stairway. Also called Smokeproof Stairway or Smokeproof Enclosure.

Sounding
Process of testing the structural integrity of a roof or floor of a building, or of locating underlying supporting members, by striking the surface of the assembly with the blunt end of a hand tool.

Stack Effect
Tendency of any vertical shaft within a tall building to act as a chimney or "smokestack" by channeling heat, smoke, and other products of combustion upward due to convection. Also called Chimney Effect.

Staging
When used in high-rise fire fighting, the Incident Command System (ICS) term for the area within the building where relief crews are assembled and spare equipment is stockpiled, usually established two floors below the fire floor. Staging may also include a first-aid station and Rehab.

Stem Wall
In platform frame construction, that exterior wall between the foundation and the first floor of a building.

Stratification
Settling at various vertical levels of accumulations or layers of smoke according to density of weight, the heaviest on the bottom.

Strip Vent
See Trench Ventilation.

Stud
Vertical structural member within a wall; most often made of wood, but some are made of light metal.

Substrate
Layer of material between a roof deck and the roof covering that may or may not be bonded to the roof covering. The most common substrate is roofing felt or tar paper.

Suspended Ceiling
Very common ceiling system composed of a metal framework suspended by wires from the underside of the roof or the floor above. The framework supports fiberboard panels that constitute the finish of the ceiling. Common applications are in office buildings and in common areas of apartment buildings and hotels.

T

Tar And Gravel Roof
See Built-Up Roof.

Tension
Those vertical or horizontal stresses that tend to pull things apart; for example, the force exerted on the bottom chord of a truss.

Thermal Layering
(1) Tendency of gases to form into layers according to temperature. (2) Process of burning in a confined space by which the hottest air is found at the ceiling and the coolest air at floor level.

Tilt-Slab Construction
Type of construction in which concrete wall sections (slabs) are cast on the concrete floor of the building and are then tilted up into the vertical position.

Top Ventilation
See Vertical Ventilation.

Trench Ventilation
Defensive tactic that involves cutting an exit opening in the roof, extending from one outside wall to the other, to create an opening at which a spreading fire may be cut off. Also called Strip Vent.

Truss
Structural assembly consisting of a top chord and a bottom chord that are separated by some triangulated configuration of web members, often found in lightweight roof construction.

UBC
Uniform Building Code published by the International Conference of Building Officials.

Vent Group
Incident Command System (ICS) term for those firefighters assigned to ventilate a structure.

Ventilation
Systematic removal of heated air, smoke, or other airborne contaminants from a structure and their replacement with a supply of fresher air.

Venturi Principle
When a fluid, such as air, is forced under pressure through a restricted orifice, there is a decrease in the pressure exerted against the sides of the constriction and a corresponding increase in the velocity of the fluid. Because the surrounding air is under greater pressure, it rushes into the area of lower pressure.

Vertical Ventilation
Ventilating at the highest point of a building through existing or created openings and channeling the contaminated atmosphere vertically within the structure and out the top. Also called Top Ventilation.

Wallboard
Fire-resistive building material consisting of a layer of highly compacted gypsum material sandwiched between two layers of paper. Also called Drywall, Plasterboard, or Sheetrock®.

Western Frame Construction
See Platform Frame Construction.

Windward
Direction from which the wind is blowing; the unprotected side.

Wired Glass
Glass in which a wire mesh is embedded during manufacture. Usually translucent, wired glass is installed in doors, windows, and skylights to increase fire resistance and security.

Introduction

VENTILATION — DEFINITION AND OBJECTIVES

Ventilation is the systematic removal of heated air, smoke, or other airborne contaminants from a structure and their replacement with a supply of fresher air. This procedure facilitates entry by firefighters for rescue and other fire fighting operations.

The importance of ventilation is obvious to firefighters because it increases visibility, which allows them to more quickly locate the seat of the fire, and because it decreases the danger to trapped occupants by channeling away hot, toxic gases. In addition, it reduces the chances of rollover, flashover, and backdraft. However, to the public its importance may not always be so apparent. Ventilation is often misunderstood by the public because it sometimes requires doing a certain amount of what may appear to be needless damage to a building; nevertheless, ventilation greatly reduces fire damage when done correctly. Ventilation activities must not be delayed nor minimized in the mistaken hope that such activities will not be required and that damage can be avoided.

Adequate ventilation is vital for both rescue and fire suppression. If the building is structurally safe, the objective is to go inside, make needed rescues, limit fire spread, and extinguish the fire. Proper ventilation cannot be accomplished haphazardly and should not be attempted independently of a coordinated fire attack as certain technical principles are involved. Firefighters should not rely solely upon knowledge gained from practical experience. They should learn ventilation theory, principles of building construction, and the techniques of making the necessary openings in fire buildings to remove accumulated combustion products.

Typically fire departments do too little ventilation rather than too much. This negative practice and the negative attitude toward ventilation must be eliminated. Properly performed ventilation can result in the saving of lives, a reduction in property damage, and safer and faster access by firefighters for more efficient fireground operations. Firefighters should be trained to begin adequate and proper ventilation activities as part of a coordinated fire attack without hesitation and without fear of criticism from superiors or the public. Fire suppression personnel must recognize that hoselines alone cannot provide efficient rescue and interior fire suppression. Prompt, adequate ventilation is the key to getting the firefighters inside the building quickly. If they cannot get inside, they need ventilation. If they have ventilated and still cannot get inside, they may need to consider even more ventilation!

Additionally, the effects of exposure to products of combustion through inhalation and absorption indicate the need for early ventilation. Since the middle of the twentieth century, the use of synthetic materials has had phenomenal growth. These materials perform so vital a function that the synthetics industry has taken its place beside the industries using wood, metal, and textiles. As a result, the fuel load in all occupancies has increased, and an increase in the toxic products of combustion also can be expected.

Knowledge of ventilation theory and practice is of paramount importance and should be applied as an integral part of fireground operations. A fire officer with an understanding of what is taking place in a fire building — and what effect certain actions will produce — is much better prepared to assume responsibility for performing ventilation.

In order to function capably as team members, firefighters need to understand the characteristics and behavior of those elements and situations with which they must contend. It is not sufficient that ventilation be understood only by the officer in charge of a fire. There are numerous instances during every fire in which each firefighter must operate independently within the tactical objectives established by the incident commander. If the firefighter has actively participated in a well-balanced, well-organized training program that includes all aspects of the profession, the firefighter will select the choice of alternatives that experience and training have taught is appropriate for the situation.

When a fire breaks out, the major objectives of a fire department are to reach the scene of a fire as quickly as safely possible, rescue trapped victims, locate the fire, and apply suitable extinguishing agents in a manner that will minimize fire, water, smoke, and heat damage. Appropriate ventilation during fire fighting is definitely an aid to the achievement of these objectives.

To achieve some of these objectives, however, it often becomes necessary to enter a structure. Although fire ventilation provides clearer conditions within a structure, it does not remove all hazards and dangerous gases. The application of water into a heavily charged, heated area may collect carbon particles from the smoke and disperse smoke and gases from the area, but it still does not eliminate the need for respiratory protection. This is the reason why all firefighters must be equipped with self-contained breathing apparatus (SCBA) during all fire fighting operations.

While ventilation is essential to rescue and fire control, there are other factors and conditions that also influence the success or failure of an operation. Failure to provide for all conditions as they arise may jeopardize an entire operation. Therefore, it is critically important that firefighters understand the principles of ventilation, its capabilities and limitations, its affects on fire behavior, and its uses in assisting in fire extinguishment.

PURPOSE AND SCOPE

The purpose of this manual is to present the principles and practices of ventilation in a manner that provides a basis for training and will enable the discerning student to learn the skills necessary to effectively contribute to reducing the loss of life and property by fire. The scope of this manual is limited to describing and illustrating the following: safe operations related to ventilation, products of combustion, elements and situations that influence the ventilation process, ventilation methods and procedures, and tools and mechanized equipment used in ventilation.

This manual is intended to be used as a reference in formal training courses on fire ventilation and in self-study by individual firefighters. To enhance the learning value of this manual — and to keep firefighter safety ever at the forefront — the applicable safety considerations are listed separately at the beginning of each chapter. The manual has been written to be consistent with the **Essentials of Fire Fighting** manual but covers the theory and practice of ventilation in greater depth. However, this manual does not attempt to cover the use of hand tools and other equipment except as necessary to safely execute ventilation assignments on the fireground. The firefighter needing more information on topics not directly related to ventilation should refer to other IFSTA manuals on the specific topic.

1

Fire Behavior And Airflow Characteristics

This chapter provides information that addresses the following performance objectives of NFPA 1001, *Standard for Fire Fighter Professional Qualifications* (1992):

Chapter 3 — Fire Fighter I

3-9.1 Define the principles of ventilation, and identify the advantages and effects of proper ventilation.

3-9.4 Identify the signs, causes and effects of backdraft explosions.

3-9.5 Identify methods of preventing a backdraft explosion.

Safety Points

In its discussion of fire behavior and airflow characteristics, this chapter addresses the following safety points:

- In order to perform ventilation and other fireground functions safely, firefighters must understand how fire and the products of combustion behave under a variety of conditions.
- Firefighters must understand the causes and effects of rollovers, flashovers, and backdrafts, as well as how to prevent or mitigate them.
- To avoid being in the path of the flame front during a rollover, firefighters should stay as low as possible when advancing hoselines within a building.
- Firefighters should know the toxic effects of the most common products of combustion and should recognize the importance of wearing SCBA during fire fighting operations, including ventilation and overhaul.

Chapter 1
Fire Behavior And Airflow Characteristics

INTRODUCTION

To function effectively as a member of a ventilation team, a firefighter must have a basic understanding of how fire, smoke, and gases behave under various conditions. This chapter discusses the basics of fire behavior, the movement of heat and smoke within confined spaces, the nature of gases encountered in both fire and nonfire situations, and the basic physics of airflow.

PHASES OF FIRE

Fire may start at any time of the day or night if the right conditions exist. If the fire occurs when the area is occupied, the chances are that it will be discovered and controlled in the incipient or beginning phase. But if it occurs when the building is closed and deserted, the fire may go undetected until it has gained major proportions (Figure 1.1). For ventilation to be performed safely and effectively, it is of primary importance that firefighters know the condition/phase of a fire within a closed building.

Figure 1.1 A fire that went undetected until it was noticed by a passing motorist.

When fire is confined in a building or room, a situation develops that requires carefully calculated and executed ventilation procedures if further damage is to be prevented and danger reduced. This type of fire can best be understood by an investigation of its three progressive stages: incipient (beginning) phase, steady-state (free-burning) phase, and hot-smoldering phase.

A firefighter may be confronted by any of the three phases of fire, and the ventilation tactics used will vary depending on the phase in which the fire is found. Firefighters must also be aware of the variety of potentially hazardous conditions — rollover, flashover, and backdraft — that may be found within the three phases. Therefore, a working knowledge of these phases is important for an understanding of ventilation procedures.

Incipient (Beginning) Phase

The incipient phase is the earliest phase of a fire, beginning with the actual ignition (Figure 1.2). The fire is limited to the original materials of ignition. In the incipient phase, the oxygen content in the air has not been significantly reduced; therefore, the fire is producing water vapor (H_2O), carbon dioxide (CO_2), and perhaps small quantities of sulfur dioxide (SO_2), carbon monoxide (CO), and other gases. Some heat is being generated, and the amount will increase with the progress of the fire. The fire may be producing a flame temperature well

Figure 1.2 Incipient phase.

above 1,000°F (537°C), yet the temperature in the room at this stage may be only slightly increased. At this point, simply opening a window in the fire room (horizontal ventilation) may be all that is required to adequately ventilate the room. At other times in the incipient phase, mechanical or forced ventilation may be required.

Steady-State (Free-Burning) Phase

For purposes of simplicity, the steady-state (free-burning) phase can generally be considered as the phase of the fire where sufficient oxygen and fuel are available for fire growth and open burning to a point where total involvement is possible (Figure 1.3). During the early portions of this phase, oxygen-rich air is drawn into the flame as convection (the rise of heated gases) carries the heat to the uppermost regions of the confined area. The heated gases spread laterally from the top and then downward, forcing the cooler air to lower levels and eventually igniting all the combustible material in the upper levels of the room. This early portion of the steady-state burning phase is often called the *flame-spread* phase. At this point, the temperature in the upper regions can exceed 1,300°F (700°C). The presence of this heated air is one of the reasons that firefighters are taught to keep low and use self-contained breathing apparatus (SCBA). One breath of this superheated air can sear the lungs.

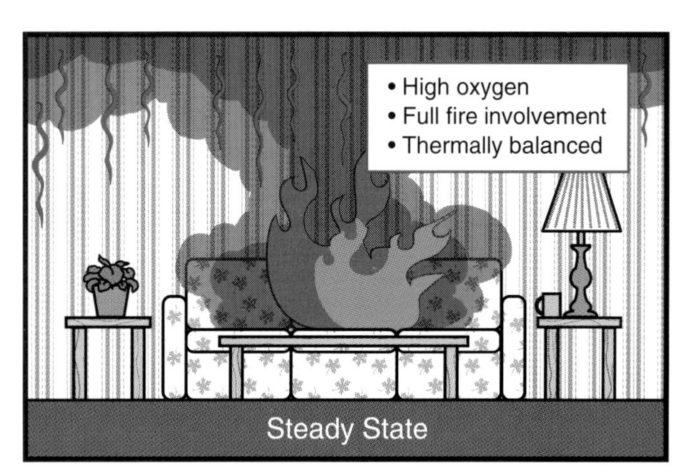

Figure 1.3 Steady-state phase.

ROLLOVER

Rollover takes place when unburned combustible gases released during the incipient and early steady-state burning phases of the fire accumulate at the ceiling level. These superheated gases are pushed, under pressure, away from the fire and into uninvolved areas where they mix with oxygen. When their flammable range is reached, they ignite; and a fire front develops, expands very rapidly, and rolls across the ceiling (Figures 1.4 a and b). This situation is why firefighters must stay low when advancing hoselines. Rollover differs from flashover (see following section) in that only the gases at the upper levels are burning — not the entire contents of the room. The rollover will continue until the fire stops producing the flammable gases that are feeding the rollover. Extinguishing the main body of the fire is the most direct way of eliminating these gases.

NOTE: Even though a hoseline is operating, rollover can occur if a sufficient volume of superheated gases are pushing into an area. Proper ventilation on the side of the fire opposite the fire attack will reduce the chances of a rollover.

Figure 1.4a Prerollover condition.

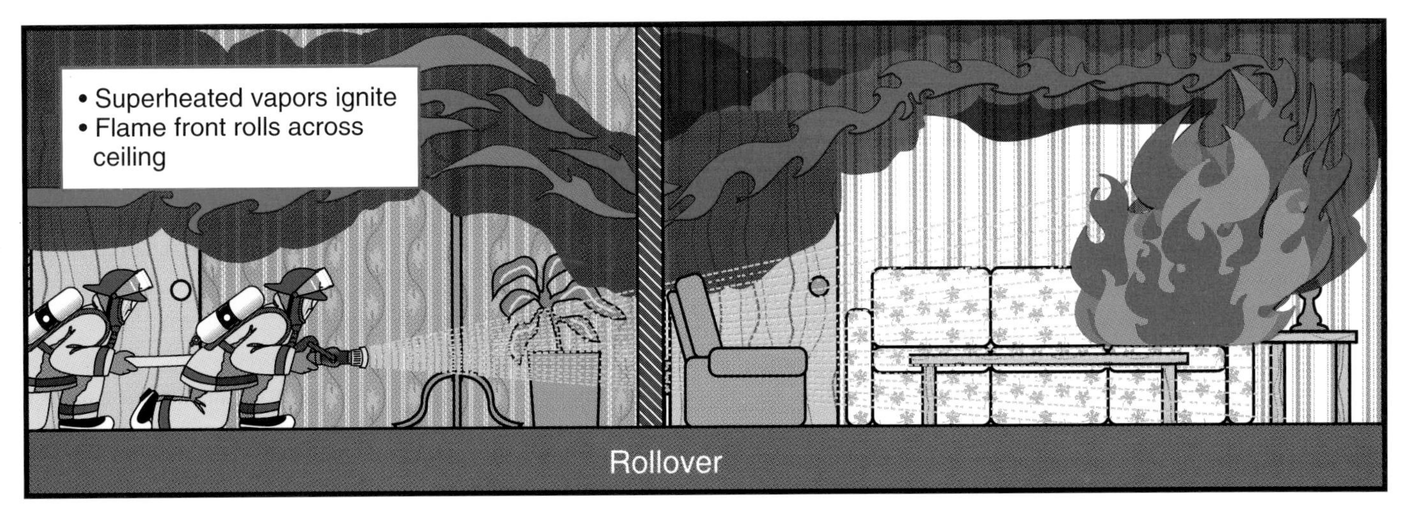

Figure 1.4b Rollover.

FLASHOVER

Flashover occurs when there is simultaneous ignition over the entire surface of a room and its contents. Originally, it was believed that flashover was caused by combustible gases released during the early stages of a fire. It was thought that these gases collected at the ceiling level and mixed with air until they reached their flammable range and then suddenly ignited causing flashover. It is now believed that while this condition may occur, as in a rollover, it precedes flashover.

The actual cause of flashover is attributed to the buildup of heat from the fire itself (Figure 1.5). As the fire continues to burn, all of the contents in the fire area are gradually heated to their ignition temperatures. When they reach this point, simultaneous ignition occurs, and the area becomes fully involved in fire. The actual ignition is almost instantaneous and can be quite dramatic. A flashover can usually be avoided by directing water toward both the ceiling level and the room contents to cool materials in the area safely below their ignition temperature and by early and aggressive ventilation.

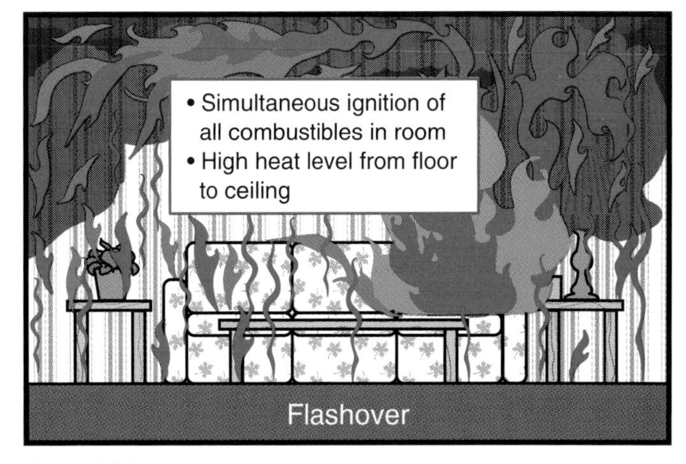

Figure 1.5 Flashover.

Hot-Smoldering Phase

As the fire progresses through the latter stages of the steady-state burning phase, it continues to consume available oxygen until it reaches the point where there is insufficient oxygen to react with the fuel (Figure 1.6). The fire is gradually reduced to the hot-smoldering phase and then needs only a new supply of oxygen to burn rapidly or explode. Proper ventilation at this point will allow the gases to be channeled harmlessly to the outside, allow hoselines to be advanced through the involved areas, and permit the remaining products of combustion to be forced out ahead of the hoselines.

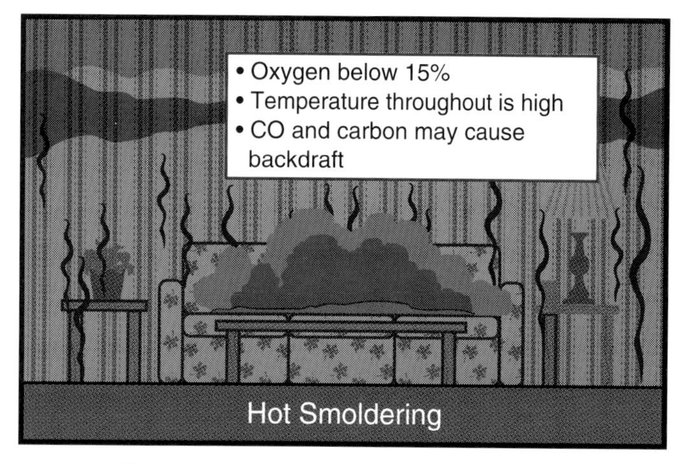

Figure 1.6 The hot-smoldering phase.

In this third phase, flame may cease to exist if the area of confinement is sufficiently airtight. In this instance, burning is reduced to glowing embers. The room becomes completely filled with dense smoke and gases to the extent that the smoke is forced under pressure from all cracks. The fire continues to smolder, and the room completely fills with dense smoke and gases at a temperature of well over 1,000°F (537°C). The intense heat continues to vaporize the lighter fuel fractions, such as hydrogen and methane, from the combustible material in the room. Because there is little available oxygen, the CO produced cannot combine with oxygen to form CO_2 as it does in the earlier phases. These fuel gases add to those produced by the fire earlier and further increase the hazard to firefighters because of the possibility of backdraft (see following section). It is critical that proper ventilation techniques be applied, allowing these explosive gases to escape harmlessly to the outside.

BACKDRAFT

Firefighters responding to a confined fire that is late in the steady-state (free-burning) phase or in the hot-smoldering phase risk causing a backdraft or smoke explosion if the science of fire is not considered before opening the structure. A *backdraft* is the rapid, almost instantaneous combustion of flammable gases, carbon particles, and tar balloons emitted by materials that are burning under conditions of insufficient oxygen.

Except under carefully controlled conditions, combustion is rarely complete. Some of the fuel elements of the burning materials are not consumed but are passed into the surrounding atmosphere in the form of unburned combustible gases. It is not necessary that a material be aflame to emit these substances. Many combustibles begin to smoke before they actually burst into flame. In such cases, either there is not enough available oxygen to support fire or the ignition temperature of the material has not been reached. Nevertheless, the gases and carbon being emitted are flammable. Proper ventilation releases smoke and hot, unburned gases from the upper areas of the room or structure. Improper ventilation at this time, such as opening a door or window (Figure 1.7), supplies the missing link — oxygen. As soon as the needed oxygen rushes in, the stalled combustion resumes with devastating speed — a backdraft occurs — truly qualifying as an explosion (Figure 1.8).

Oxidation is a chemical reaction in which oxygen combines with other elements, and *combustion* is merely rapid oxidation. A backdraft is *extremely* rapid oxidation, on the order of an explosion. Carbon is a naturally abundant element present in wood, plastics, petroleum products, and all other organic materials. When these materials burn in the presence of abundant oxygen, the CO produced combines with the available oxygen to form CO_2. However, if the available oxygen is limited, some or all of the carbon monoxide has no oxygen with which to unite. Under these conditions, free carbon is released in the smoke. A warning sign of possible backdraft is this dense, black, carbon-filled smoke in a confined space. When oxygen is reintroduced into the space, it combines with the carbon, and the smoke becomes less black, turning yellow or grayish yellow instead.

Figure 1.7 Prebackdraft condition.

Figure 1.8 Backdraft.

The following characteristics may indicate a backdraft or smoke explosion condition:

- Smoke under pressure exiting small openings
- Black smoke becoming dense, grayish yellow
- Confinement and excessive heat
- Little or no visible flame
- Smoke leaving the building in puffs and being drawn back in
- Smoke-stained and/or rattling windows
- Muffled sounds
- Sudden, rapid movement of air and smoke inward when an opening is made

Upon arrival at a fire, firefighters must exercise extreme caution if the weather boards or other outside wall materials are hot, or even noticeably warm, and little or no fire is evident (Figure 1.9). These signs indicate that a hot, smoldering fire may have been burning in the building for a considerable length of time, giving off large quantities of gases that are still within the building. This assumes that the building is still tightly closed and that the surface heat is not caused by fire inside the wall where the heat is detected.

The sound of building windows rattling (very much like in a heavy windstorm) is usually an indication that backdraft conditions exist (Figure 1.10). The actual cause of this rattling sound is not definitely known, but experience has shown that this phenomenon is a definite warning sign of backdraft conditions. Without proper ventilation, an explosion can be expected when oxygen is reintroduced.

Figure 1.9 A firefighter feels an exterior door for heat.

Figure 1.10 A firefighter listening for a rattling window.

If visibility through the windows is practically nil because of smoke but no fire is in evidence (or perhaps only a faint orange glow can be seen through the smoke), it is reasonable to assume that the fire has been smoldering for some time in order to produce sufficient smoke to cause such poor visibility. Extreme caution must be exercised because danger exists, and horizontal ventilation should not be attempted if any other options are available.

Assuming the conditions are recognized upon arrival, every effort must be made to prevent the backdraft from occurring. Prompt and proper ventilation is in order, but even this procedure cannot be relied upon to absolutely prevent the backdraft. In fact, there is no sure, guaranteed method of prevention.

In the interest of safety, personnel should not stand in doorways nor directly in front of doors, windows, or other openings when the possibility of a backdraft exists. It is safest if they do not stand within the V-shaped force pattern that would emanate from such openings (Figure 1.11). The gaseous products of the backdraft will expand as they come through the opening because of the lesser pressure of the atmosphere outside the building.

Figure 1.11 The V-shaped danger zone outside of a doorway.

It is possible that backdraft conditions may not exist upon arrival but develop in confined spaces within a building while firefighters are in the building. These conditions may develop in spaces remote from the original fire area. If firefighters overlook the warning signs in their haste to get into the building and extinguish the fire, they may be inside the building when backdraft conditions begin to develop.

Fortunately, there are also warnings that are discernible inside the building. Experienced firefighters know the usual appearance of fire, and they may describe the color of the flames as orange-yellow or reddish orange and say that the flames appear lively. However, flames assuming a pale or sickly yellow hue and seeming to lose their liveliness (appearing somewhat like a slow-motion picture) inside a confined space within the building is an indication that the fire is becoming oxygen-starved and that unburned flammable gases are being emitted. Firefighters may also become aware that sounds — such as voices, personal alert safety system (PASS) devices, and the noises emitted by the fire — have become unusual and somewhat muffled.

A further development of this condition becomes noticeable when the fire on which no water has yet been applied diminishes in intensity or when the flame subsides almost completely. This condition indicates that the amount of oxygen in the atmosphere is extremely low and that backdraft conditions may be developing.

If firefighters are inside a building when backdraft conditions begin to develop, they should leave as quickly as possible. If it is too late for them to leave, they should drop to the floor (Figure 1.12). While the force of the explosion may be evident from floor to ceiling, the fire itself may only come to within a few feet of the floor. Firefighters who stay low have the best chance of not being burned or thrown about by the force.

While vertical ventilation is usually the best solution to the problem of safely mitigating backdraft conditions, it may not always be possible. Each enclosed section of the building must be treated as a separate potential backdraft area (Figure 1.13). If the confined space is located below the top floor of a multistory building, it may not be practical to attempt to ventilate it vertically. If it is

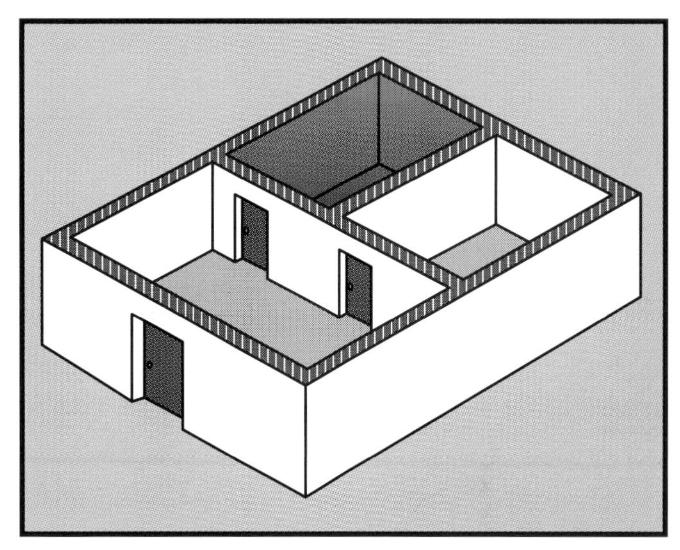

Figure 1.13 Backdraft conditions can develop in any tightly sealed space.

decided that the space cannot be allowed to cool down before making entry, horizontal ventilation may be the only available option under these conditions. This ventilation must be performed in such a way as to maximize firefighter safety and minimize property damage (Figure 1.14).

Once backdraft conditions develop within a confined space, firefighters have only a few safe courses of action. Under these conditions, it may be necessary simply to do nothing except to monitor

Figure 1.12 Fire passing over firefighters on the floor.

Figure 1.14 Horizontal ventilation procedures must maximize firefighter safety and minimize property damage.

the space until the smoldering fire goes out because of a lack of oxygen or fuel and to allow the gases within the space to cool below their ignition temperature before opening the space. This procedure could be very time consuming, so it is often not a practical solution to the problem. Another course of action may be to allow the backdraft to take place but to control its path by ventilating vertically so that the energy is released harmlessly into the atmosphere. The use of a piercing nozzle to penetrate the space and accelerate the cooling process by initiating a form of indirect fire fighting may also be an option (Figure 1.15).

Figure 1.15 Firefighters about to penetrate a wall with a piercing nozzle.

PRODUCTS OF COMBUSTION

When a material (fuel) burns, it undergoes a chemical change. None of the elements making up the material are destroyed in the process, but all of the matter is transformed into another form or state. The products of combustion have the same weight as the fuel had before it was burned. These products of combustion are forced, under pressure, up and away from the fire area by their *buoyancy* (tendency to rise or float) and greatly expanded volume created by a rise in temperature. When a fuel burns, there are four products of combustion: flame, heat, smoke, and fire gases (Figure 1.16).

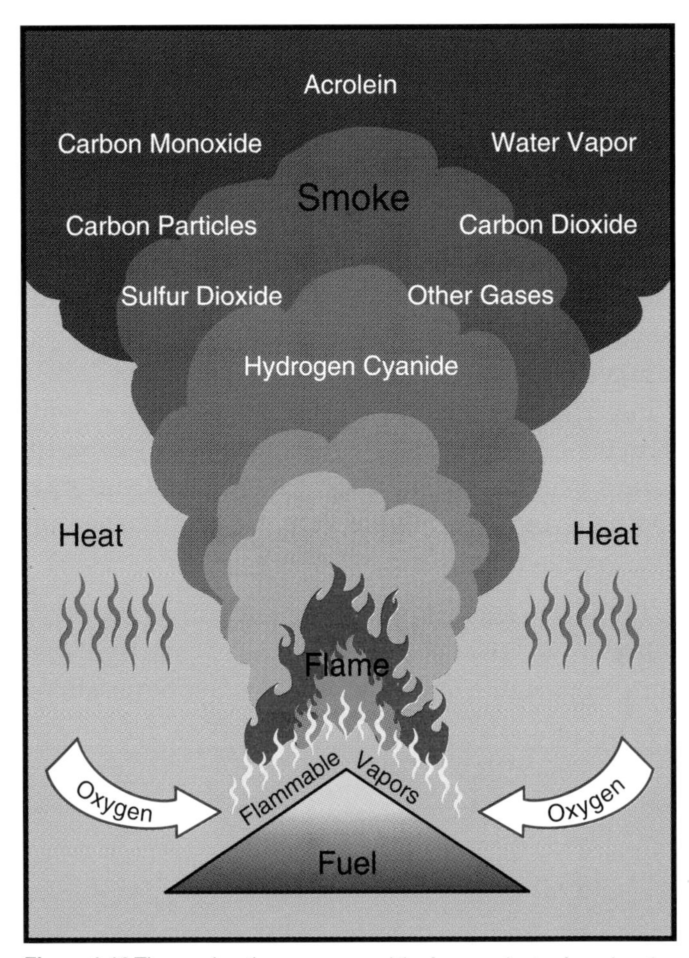

Figure 1.16 The combustion process and the four products of combustion (flame, heat, smoke, and fire gases).

Flame

Flame is the visible, luminous body of burning gas. When a burning gas is mixed with the proper amounts of oxygen, the flame becomes hotter and less luminous. This loss of luminosity is caused by a more complete combustion of the carbon. For these reasons, flame is considered to be a product of combustion. However, heat, smoke, and gas can develop in certain types of smoldering fires without evidence of flame.

Heat

Heat is a form of energy that is measured in British thermal units (Btu) or in calories (cal) or, in the International System of Units (SI), by joules (J). A *Btu* is the amount of heat required to raise the temperature of 1 pound of water 1 degree Fahrenheit. A *calorie* is the amount of heat required to raise the temperature of 1 gram of water 1 degree Celsius. A *joule* is the amount of heat energy provided by 1 watt flowing for 1 second. A Btu equals 1,054 joules, and 1 calorie equals 4.183 joules. Neither Btu nor calories are approved SI units, but joules are.

Temperature is associated, and sometimes confused, with heat. The temperature of a material is the condition that determines whether it will transfer heat to or from other materials. Therefore, *temperature* is an *indicator* of heat. Temperature is measured in degrees Fahrenheit (F) or in SI by degrees Celsius (C).

Because the existence of heat within a substance is caused by molecular motion, the greater the molecular activity, the more intense the heat and the higher the temperature. In a physiological sense, heat is a direct cause of burns and other forms of personal injury such as dehydration, heat exhaustion, and injury to the respiratory tract.

A number of natural laws of physics are involved in the transmission of heat. One is called the Law of Heat Flow, which specifies that heat tends to flow from a warmer to a cooler substance. The colder of two bodies in contact will absorb heat until both objects are the same temperature. Heat can travel throughout a burning building by one or more of three methods. These methods are commonly referred to as conduction, radiation, and convection. These phenomena are discussed later in this chapter.

Smoke

Firefighters should expect a closed building in which a fire is burning to be filled with smoke and hot gases. These products are dangerous to the physical welfare of those entering the building. In addition, these smoke and hot gases make it difficult for firefighters to move through the building to locate the fire.

Most of the smoke at a fire is a suspension of small particles of carbon and tar, but there are also some ordinary dust and other particulates floating in a combination of heated gases. The particles provide a means for the condensation of some of the gaseous products of combustion, especially aldehydes and organic acids formed from carbon. Some of the suspended particles in smoke are merely irritating, but others may be lethal. The size of a particle determines how deeply into the unprotected lungs it can be inhaled.

One of the airborne particulates that may be present in smoke is asbestos. Asbestos was widely used for many years as a fire-resistant insulation around pipes and as a coating on ceilings. Although generally safe when left undisturbed, asbestos becomes extremely dangerous when airborne because of its long-term carcinogenic potential. One way to reduce the hazard of disturbed asbestos insulation is to continually wet down any suspected asbestos-containing material to prevent it from becoming airborne. Once again, it is obvious why self-contained breathing apparatus should always be worn during ventilation operations. While SCBA provides some protection from airborne asbestos and other insulating materials, it must also be remembered that asbestos-contaminated protective clothing can also be a significant hazard if not properly handled.

Fire Gases

The volume of the various gases produced in most fires is well in excess of the amount necessary to be considered lethal. Many gases found in fire buildings have a vapor density greater than 1.0, meaning they are heavier than air at *normal* temperatures and would *normally* be found near the floor. However, because gases become *lighter* as they are heated, they tend to expand and rise. Most of the gases encountered in a fire building are hot because they are products of combustion; thus, the greatest concentration of these gases is likely to be located in the upper portions of a fire room. If the fire is extinguished without complete ventilation, the heavy, heated gases will cool and return to the lower levels during overhaul. Therefore, the continued use of protective breathing apparatus is necessary (Figure 1.17).

Figure 1.17 Firefighters should continue to use SCBA during overhaul.

Entering a fire area means exposure to a combination of irritants and toxicants that cannot be accurately predicted. In fact, the combination can have a synergistic effect in which the combined effect of two or more substances is more toxic or more irritating than the total effect of each being inhaled or absorbed separately.

Inhaled toxic gases may have several harmful effects on the human body. Some of the gases damage lung tissue and impair its function. Other toxic gases have no harmful effect on the lungs directly but pass into the bloodstream and to other parts of the body.

The particular toxic gases given off at a fire vary according to the following four factors:

- The nature of the combustible
- The rate of heating
- The temperature of the evolved gases
- The oxygen concentration

It would be impossible to list all fire gases that exhibit toxic effects. Also, it would be impractical because the amount present and the doses inhaled, ingested, or absorbed vary so widely. Therefore, no attempt has been made to record the exact dose necessary to cause harmful effects. The gases discussed in the following sections are the ones most commonly produced during a fire and are those for which timely and proper ventilation is required. For specific information on the physiological effects of these gases, refer to the IFSTA **Self-Contained Breathing Apparatus** manual.

CARBON MONOXIDE

More fire deaths are caused by carbon monoxide (CO) than by any other toxic product of combustion. This colorless, tasteless, and odorless gas is present with every fire. The poorer the supply of oxygen, the more incomplete the burning and the greater the quantity of carbon monoxide liberated. Although subject to much variation, a good rule of thumb is: the darker the smoke, the higher the carbon monoxide level. Black smoke is high in particulate carbon and carbon monoxide because of incomplete combustion.

Carbon monoxide does not act directly on the body but crowds oxygen from the blood. The blood's hemoglobin combines with and carries oxygen in a loose chemical combination called *oxyhemoglobin*. Because of hemoglobin's affinity for carbon monoxide, it combines with CO about 200 times more readily than with oxygen, so the available oxygen is excluded. Thus, the loose combination of oxyhemoglobin becomes a strong combination called *carboxyhemoglobin* (COHb). This situation leads to eventual hypoxia of the brain and tissues, followed by death if the process is not reversed.

Because CO is produced in such abundance in fires in confined spaces, another compelling reason for the early and effective ventilation of these spaces is its flammability. With an extremely wide flammable range of 12.5 percent to 74 percent and a relatively low autoignition temperature of 1,128°F (609°C), CO can be one of the most dangerous and destructive gases with which firefighters are faced. It is the single largest contributor to the backdraft conditions discussed earlier in this chapter.

Even after all visible smoke has been cleared from a confined space where a fire has been burning, invisible but very toxic products of combustion can continue to be present in dangerous concentrations. In addition to the CO that will always be present after a fire, any of several other toxic gases may be present also, depending upon the type of materials that were involved. If a broad-spectrum gas analyzer is available, it is strongly recommended that the atmosphere within the structure be checked for all known or suspected gases that may be present. However, if only a CO meter is available, CO can be used as an *index gas*. The basic assumption is that if CO has been reduced to safe levels by ventilation, any other gases present will also have been similarly reduced. In order for CO to be used as an index gas, however, the meter must be capable of *measuring* the CO level — not just detecting its presence (Figure 1.18). Some fire departments require a CO meter reading of 50 parts per million (ppm) or lower before personnel are allowed to work inside a space without SCBA. Fire departments are encouraged to research this concern and define safe limits and procedures for dealing with CO and other fire gases.

Figure 1.18 A firefighter sampling interior atmosphere for CO level.

OTHER COMMON FIRE GASES

Other common fire gases and their sources are as follows:

- *Acrolein* (CH_2=CHCHO) is a strong respiratory irritant that is produced when polyethylene is heated and when materials containing cellulose, such as wood and other natural materials, smolder. It is used in the manufacture of pharmaceuticals, herbicides, and tear gas.
- *Hydrogen chloride* (HCl) is a colorless but very pungent and irritating gas given off in the thermal decomposition of materials containing chlorine such as polyvinyl chloride (PVC) and other plastics.
- *Hydrogen cyanide* (HCN) is a colorless gas with a characteristic almond odor. Twenty times more toxic than CO, it is an asphyxiant and can be absorbed through the skin. HCN is produced in the combustion of natural materials, such as wool and silk, that contain nitrogen. It is also produced when polyurethane foam and other materials that contain urea burn. The concentrated bulk chemical is also used in electroplating businesses.
- *Carbon dioxide* (CO_2) is a colorless, odorless, and nonflammable gas produced in free-burning fires. While it is nontoxic, CO_2 can asphyxiate by excluding the oxygen from a confined space. It also can increase a person's intake of toxic gases by accelerating his or her rate of breathing.
- *Nitrogen oxides* (NO_2 and NO) are two toxic and dangerous substances liberated in the combustion of pyroxylin plastics. Because nitric oxide (NO) readily converts to nitrogen dioxide (NO_2) in the presence of oxygen and moisture, nitrogen dioxide is the substance of most concern to firefighters. Nitrogen dioxide is a pulmonary irritant that can also have a delayed systemic effect. The vapors and smoke from the oxides of nitrogen have a reddish brown or copper color.
- *Phosgene* ($COCl_2$) is a highly toxic, colorless gas with a disagreeable odor of musty hay. It may be produced when refrigerants, such as

freon, contact flame. It can be expected in fires in cold-storage facilities. It may also be a factor in fires in heavy-duty heating, ventilating, and air-conditioning (HVAC) systems. It is a strong lung irritant, the full deleterious effect of which is not evident for several hours after exposure.

BEHAVIOR OF HEAT, SMOKE, AND FIRE GASES

Fire spreads in various ways because of the different means by which heat is transmitted, and the resulting smoke and fire gases also behave in different ways. In addition to the ways in which heat is transmitted, the behaviors of smoke and fire gases of most concern from a ventilation standpoint are mushrooming, thermal layering, and pressure differential.

Transmission Of Heat

Depending on a number of variables, fires inside buildings spread in different ways and at different rates. Because these fires are less affected by prevailing winds than are wildland fires, the principal variables are the ways in which heat is transmitted. The three recognized ways are conduction, radiation, and convection.

CONDUCTION

The process of conduction is generally associated with heat transfer, rather than with flame. Heat may be conducted from one body to another by their being in direct contact or by an intervening heat-conducting medium (Figure 1.19). The amount of heat that will be transferred and its rate of travel by this method depend upon the conductivity of the material through which the heat is passing. Not all materials have the same heat conductivity. Aluminum, copper, and iron are good conductors. Fibrous materials, such as wood, felt, cloth, and paper, are poor conductors.

Liquids and gases are poor conductors of heat because of the movement of their molecules. Air is a relatively poor conductor. Certain solid materials when shredded into fibers and packed into batts make good insulation because the material itself is a poor conductor and because there are air pockets within the batting.

Figure 1.19 Heat can be conducted through walls by direct contact with pipes or ducts.

RADIATION

The best example of radiation is the heat from the sun. Although air is a poor conductor, it is obvious that heat can travel where matter does not exist. This method of heat transmission is known as *radiation of heat waves*. Heat and light waves are similar in nature, but they differ in length per cycle. Heat waves are longer than light waves, and they are sometimes called *infrared rays*. Heat is radiated equally in all directions from its source and will travel through space until it strikes an opaque object. As the object is exposed to thermal radiation, it will in turn radiate heat from its surface. Radiated heat is one of the major sources of fire spread (especially to exposures) and is the primary mechanism by which flashover occurs, so the source of the radiation (the fire) should be eliminated as quickly as possible (Figure 1.20).

CONVECTION

Convection is the transfer of heat by the movement of fluids. When a fluid, such as water or air (or any other gas), is heated, it expands, becomes lighter, and begins to move. Because the heated

Figure 1.20 Fire spreading to exposures by radiation.

fluid is lighter than the surrounding medium, it will rise. As the heated air moves upward, cooler air is drawn in to take its place at the lower levels. The process can be observed simply by watching the smoke from a lighted cigarette in a draft-free area. The smoke from the cigarette rises, virtually straight up, until it is cooled by the surrounding air. In the same manner, air in contact with a steam radiator becomes heated by conduction. It expands, becomes lighter, and moves upward by convection. This generally upward movement of heated fluids is known as *heat transfer by convection*.

Heated air in a building will expand and rise. For this reason, fire spread by convection is mostly in an upward direction, although air currents can carry heat in any direction. Convected currents are generally the cause of heat movement from room to room, area to area, and floor to floor (Figure 1.21). The spread of fire through corridors, up stairwells and elevator shafts, between walls, and through attics is mostly caused by the convection of heat currents. Convection has more influence upon the positions for fire attack and ventilation than either radiation or conduction.

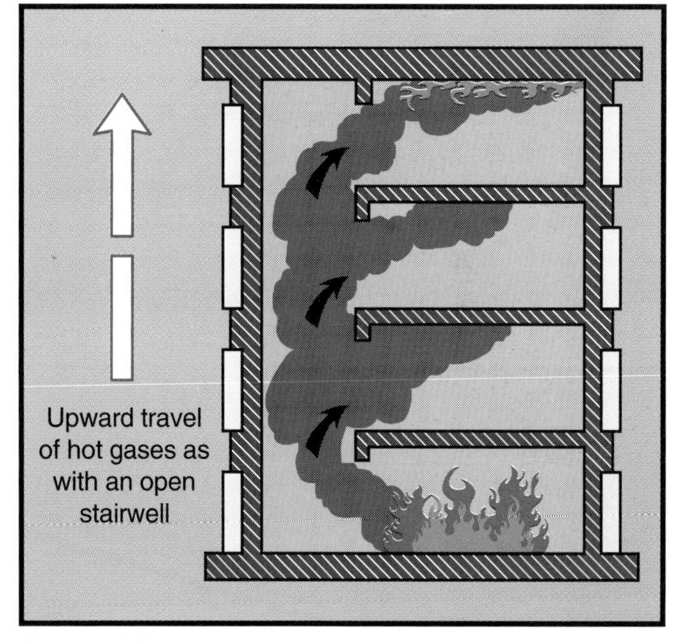

Figure 1.21 Fire spreading upward by convection.

Mushrooming

Mushrooming is caused by the fire's convection currents. Inside a confined space, smoke and fire gases rise until they reach the top of the space. Here they spread out laterally until they again encounter another obstruction. They then begin to bank down, filling the space from the top down (Figure 1.22). The accumulation of smoke, heat, and gases at these upper levels poses many dangers to occupants and firefighters, including fire spread to upper floors, backdraft, and toxicity. Vertical ventilation prevents mushrooming, particularly when it is performed over vertical shafts such as stairways.

Figure 1.22 An example of mushrooming in an ordinary dwelling.

Thermal Layering

When a fire burns in a room, thermal energy causes the surrounding air to expand. Heated fire gases collect in what is called a *hot layer*. As the fire grows in intensity, this hot layer grows as well, gradually descending from the ceiling. The temperature of the hot layer increases, the amount of unburned fuel gases increases, and the oxygen content of the hot layer is greatly reduced.

The hot layer contributes to the spread of a fire because the heat it radiates preheats the fuel beneath it (Figure 1.23). This effect can be removed through proper ventilation techniques early in the fire development stage. When fire gases are confined, heat and pressure increase. As the gases expand, they become less dense than the surrounding atmosphere and they rise. This plays an important role in creating both the hot layer and a condition called the *stack effect*, which is discussed later in this chapter.

Pressure Differential

As the heat generated from a fire rises within a confined area, pressure (slightly above atmospheric pressure) is created at the upper level of the room. Consequently, a low-pressure condition is created at the lower level of the room as air is drawn upward by the burning process. This variation in pressure is referred to as *pressure differential*. The term is also used to describe the condition created when a blower is used in a window or doorway. This effect will be explained later in this chapter.

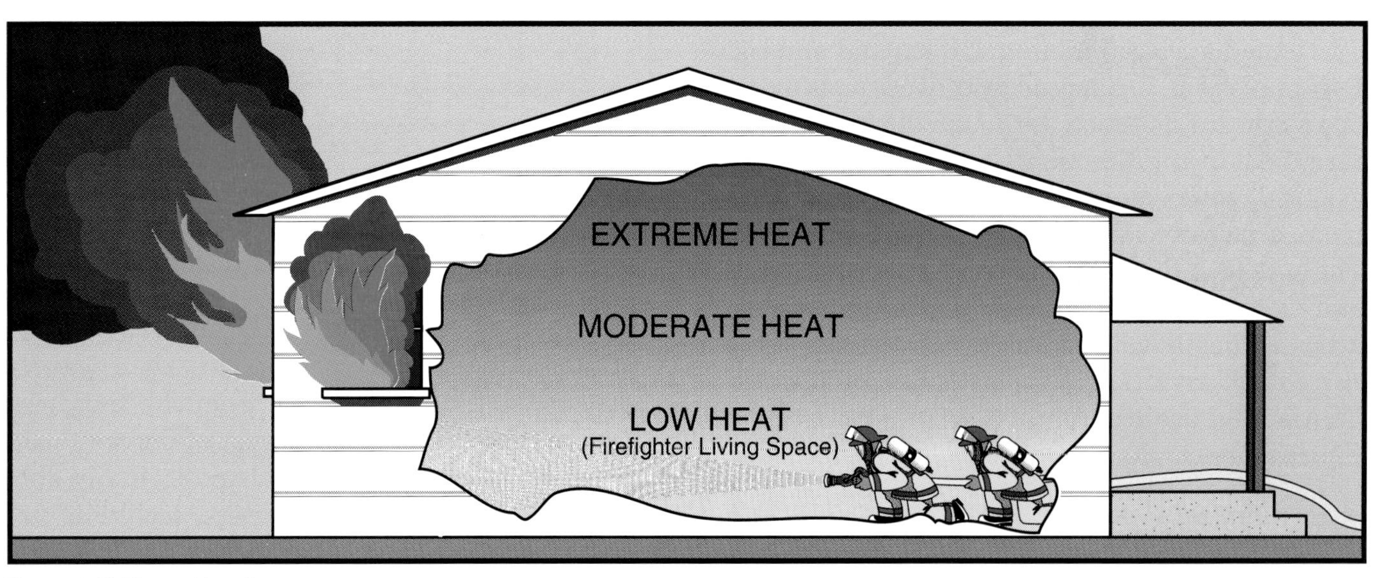

Figure 1.23 Thermal layering.

Inversions

Temperature inversions, often called *night inversions*, also affect the behavior of smoke and fire gases. During a night inversion, there is a layer of cool air at and just above ground level. The air temperature increases with elevation. This temperature difference might be as much as 25°F (-4°C) in 250 vertical feet (76.2 m). Smoke will rise until its temperature matches that of the surrounding air, and it will then stratify and extend horizontally (Figure 1.24).

NONFIRE GASES

Hazardous atmospheres can also be found in numerous situations where fire is not involved. Many industrial processes use extremely dangerous chemicals to manufacture ordinary items (Figure 1.25). For example, large quantities of carbon dioxide would be stored at a facility where dry ice or carbonated soft drinks are produced.

In addition, many common substances possess hazardous characteristics when released or combined with other materials. Refrigerants, for example, are toxic and may be accidentally released, causing a rescue situation to which firefighters may be called. Sulfur dioxide is a refrigerant that will react with moisture in the lungs to form sulfuric acid.

Figure 1.25 Chlorine used in a manufacturing process.

Rescues in sewers, caves, trenches, storage tanks, tank cars, bins, silos, manholes, pits, and other confined places require the use of self-contained breathing apparatus because of the

Figure 1.24 Smoke stratifying because of night inversion.

presence of some toxic gas or an oxygen deficiency (Figure 1.26). Workers have been overcome by harmful gases in large tanks during cleaning or repairs. Unfortunately, personnel attempting to rescue them without wearing SCBA have also been overcome. Even if there are no toxic gases present, the atmosphere in many of these tanks is oxygen deficient and will not support life. The need to use SCBA is just as important in these situations as it is in fire situations. Adequate ventilation of these areas will eliminate the hazard.

Figure 1.26 A rescuer entering a confined space.

Heavier-Than-Air Gases

Gases with a specific gravity greater than that of air are called heavier-than-air gases. These gases will gravitate to and collect in low areas unless they are heated and/or influenced by air currents.

Chlorine gas is an example of a heavier-than-air gas. Being about 2½ times heavier than air, chlorine gas will flow along the ground if released. Carbon dioxide, another heavier-than-air gas, is not considered toxic but will displace the oxygen from a confined space, so it must also be ventilated (Figure 1.27).

Figure 1.27 A space filled with carbon dioxide must also be ventilated.

Heavier-than-air gas releases will often involve a blanketing effect as the gas spreads over an area. The release can sometimes be dissipated through the use of water spray or, if indoors, ventilation devices.

Lighter-Than-Air Gases

Gases with a specific gravity less than that of air are called lighter-than-air gases. These gases will rise without being heated and without the influence of air currents. Methane, with a specific gravity of 0.554; hydrogen, with a specific gravity of 0.069; and acetylene, with a specific gravity of 0.906, are examples of lighter-than-air gases.

Because lighter-than-air gases tend to diffuse readily into the atmosphere, the frequency of potentially high concentrations is less than that with heavier-than-air gases. However, a lighter-than-air gas will also spread much more rapidly than a heavier-than-air gas, so its release must still be stopped and the area ventilated (Figure 1.28).

Figure 1.28 A firefighter directing fan toward the ceiling to ventilate lighter-than-air gases.

Measuring Toxic Atmospheres

Airborne contaminants can be either gases, vapors, or particulates. Determining in what concentrations these materials exist within a confined space can be a slow, time-consuming process. In general, when operating in confined spaces, firefighters must assume that the contaminants are present at lethal levels until it can be determined otherwise.

There are a number of different devices available to sample air for toxicity (Figure 1.29). Selecting which type of instrument to use involves a knowledge of air sampling practices and types of air contaminants. The sampling can be accomplished through area sampling or personal sampling for brief or extended periods. For most fire departments, the time involved in getting definitive results makes most of these methods impractical for use on an emergency scene. The primary exception is sampling the atmosphere for carbon monoxide as described earlier in this chapter.

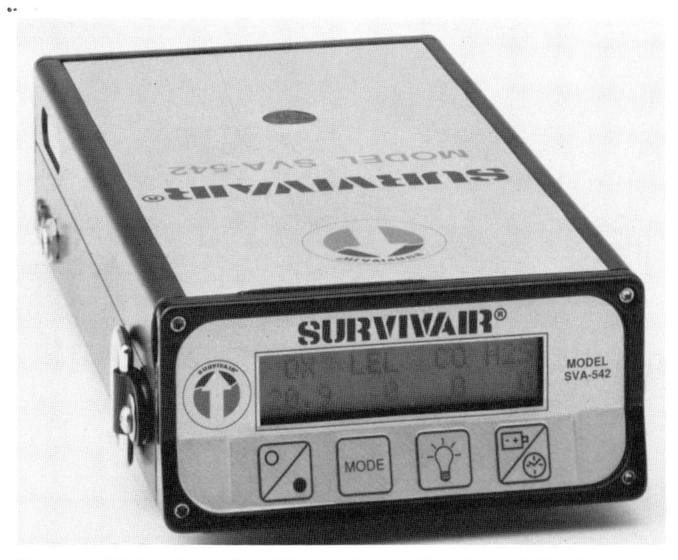

Figure 1.29 One type of multi-gas detector. *Courtesy of Survivair.*

AIRFLOW CHARACTERISTICS

For firefighters to have a more complete understanding of all aspects of the ventilation process, some understanding of the physics of airflow is necessary. This section addresses those characteristics that are the most relevant to the ventilation process.

Pressure Transfer

Just as heat flows spontaneously from a hotter object to a colder object until their temperatures are equal, pressure transfers from an area of higher pressure to another of lower pressure until the pressures are equal (Figure 1.30). If two rooms are connected by an opening and one room is pressurized, air from the pressurized room will move into the other room, equalizing the pressure across both rooms.

When gases within a structure are heated, they expand and become less dense than the surrounding atmosphere. This buoyancy causes the gases to

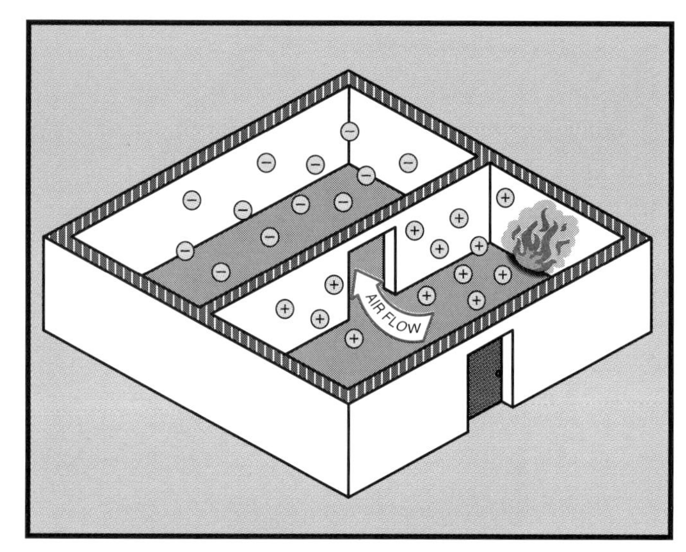

Figure 1.30 Pressure transfers from an area of higher pressure to another of lower pressure.

rise, and if they are confined, the internal pressure of the structure will increase. However, if buoyancy forces are the only ones acting on the gases within a closed compartment, then in relation to atmospheric pressure, the pressure will be higher near the top of the compartment and lower near the bottom of the compartment.

At about the vertical midpoint of a room or space, there is a plane called the *neutral pressure plane* where the interior pressure is equal to the pressure outside the space. This concept is important in understanding the phenomena of stack effect and backdraft. While the location of the neutral pressure plane can be calculated from engineering formulas, it is not practical on the fireground. However, a basic understanding of the pressure changes is needed to be able to understand certain ventilation practices, particularly those used in high-rise structures.

Assuming that the structure is closed, with little or no gases being vented to the atmosphere, a neutral pressure plane will form in the structure. This plane will be at normal atmospheric pressure. As the distance away from this plane increases, the pressure difference also increases—positively above the plane and negatively below the plane (Figure 1.31). Because there is no exchange of gases in a closed structure, upper areas of the structure will be under positive pressure, and negative pressure will form at the bottom of the structure.

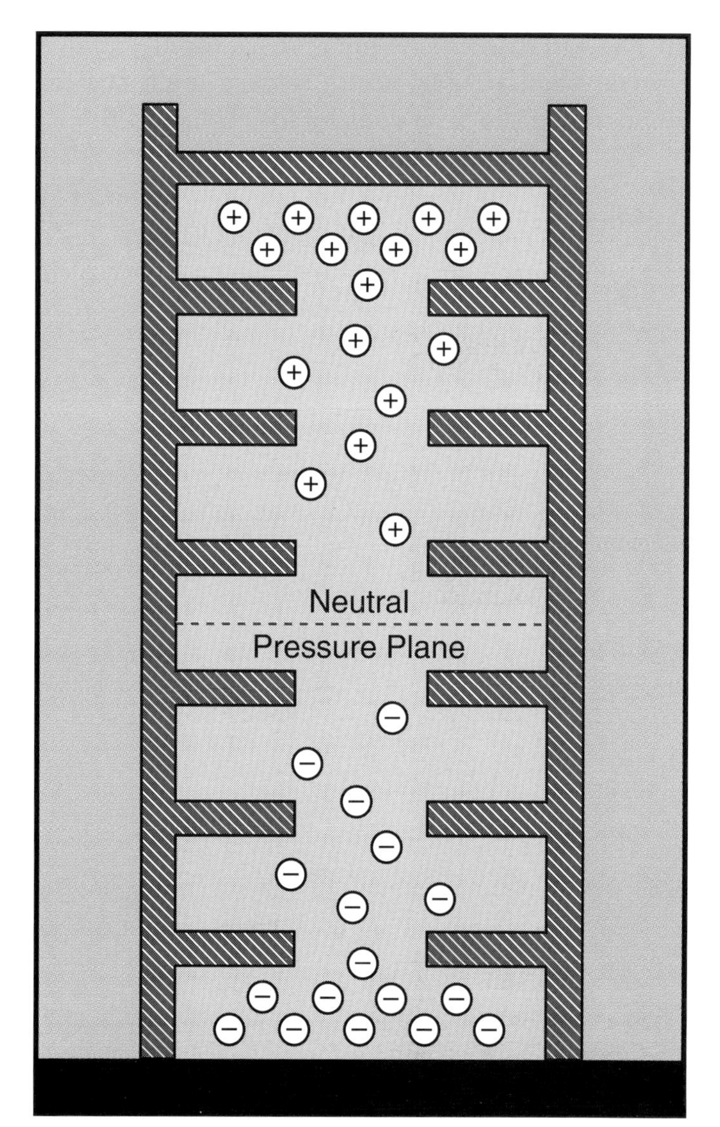

Figure 1.31 Neutral pressure plane.

This condition has a pronounced effect on tall structures, such as high-rise buildings, where the distances from the top and bottom to the neutral plane are extreme. If openings are made at the top and bottom of the structure, positive pressure located at the upper end of the structure will escape upward, and the negative pressure located at the bottom will create an inward pull of air, causing the entire structure to act as a chimney. This phenomenon is called the *stack effect* (Figure 1.32). The intensity of the stack effect is greater in situations where a significant difference in temperature exists between the inside and the outside of a building. Wind can also have a major influence on stack effect because of its ability to produce a positive pressure on the windward side of the building and a negative pressure on the leeward side.

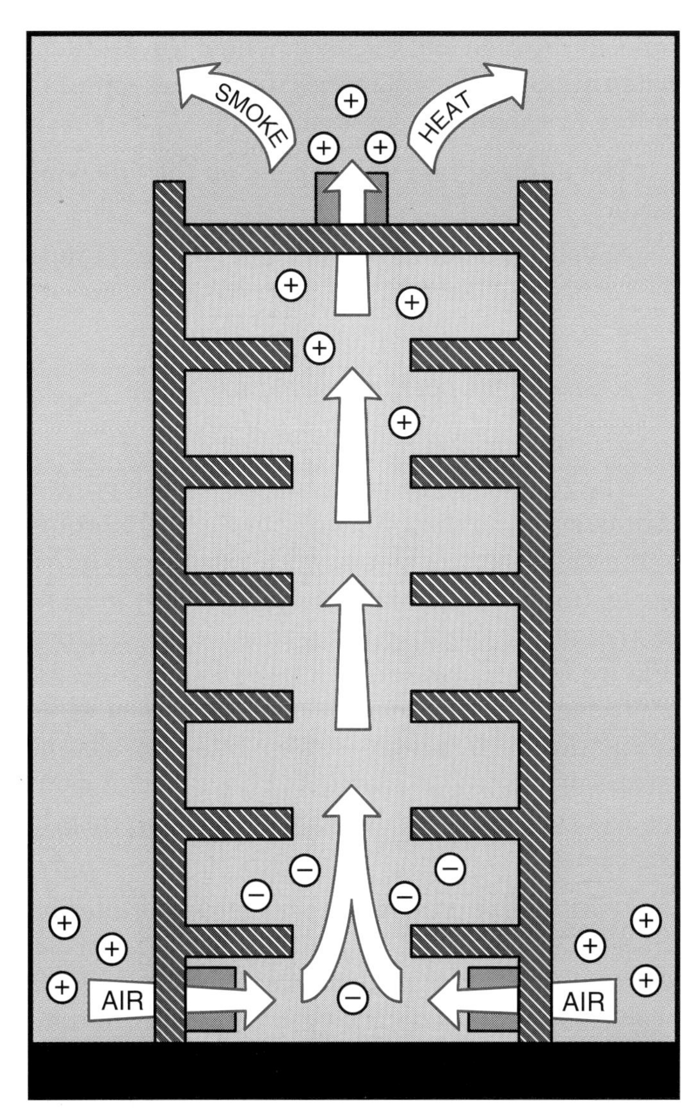

Figure 1.32 Stack effect.

The effect of the neutral pressure plane can also be seen in a backdraft situation, where there is a pressure buildup inside a nonvented building. The negative pressure at lower portions of the structure will draw air in through the narrow cracks at the bottoms of doors. When that minute amount of air mixes with the superheated combustible gases just inside the door, ignition occurs; and a puff of smoke will issue from the same crack that allowed the air in.

Mechanical ventilation creates a pressure differential in the area being ventilated. When air is removed, a low-pressure environment is created inside the structure, allowing air to enter. When air is added, a high-pressure environment is created inside the structure, causing air to move to the outside (a low-pressure area). This difference in pressure is what enables firefighters to ventilate a structure (Figure 1.33).

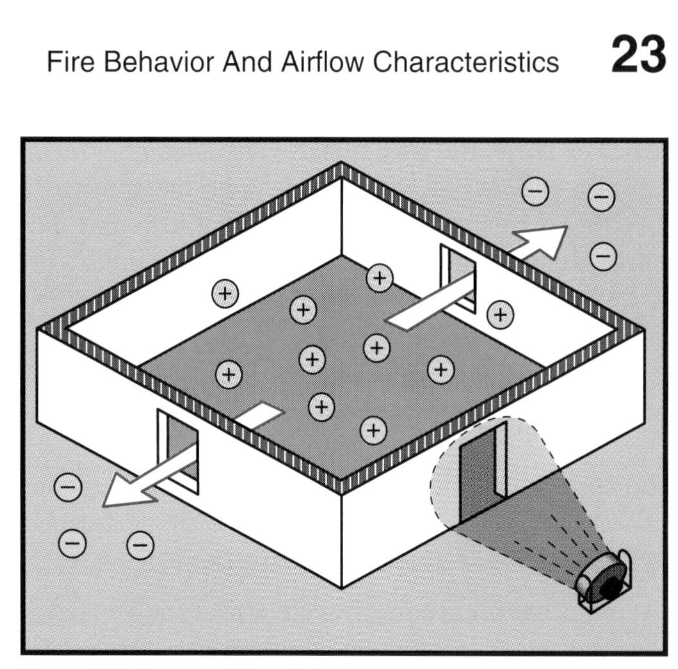

Figure 1.33 Pressure differential.

Diffusion

As with heat and pressure transfer, gas molecules move from an area of high concentration to areas of lower concentration through a process known as *diffusion*. Diffusion occurs in both liquids and solids. Diffusion is the reason we can smell a bottle of perfume shortly after it is opened, even if we are some distance away. The high concentration of perfume molecules quickly disperses throughout the room, unless influenced by air currents in the room. Diffusion may not be important in fire suppression operations, but it may be very important during nonfire ventilation operations and overhaul. Toxic gases can permeate a structure in a very short time. Without precise measurement of atmospheric samples, firefighters must assume that any toxic gas inside a structure has permeated throughout the entire room or rooms.

Circulation

Circulation is the moving around of the air inside a structure or other confined space. While circulation can be of some benefit, such as when cooling an office with a fan, it is of little benefit in reducing toxic atmospheres. By inadequately sealing the area around a smoke ejector, firefighters sometimes only circulate contaminated air instead of removing it. Only in rare cases, such as in buildings with dead air spaces, will circulation be of benefit to ventilation efforts. Because void spaces tend to trap air and resist ventilation, a large volume of fresh air may have to be mechanically circulated through these spaces to dilute and disperse the toxic gases and clear the structure.

Dilution

Controlling the movement of heat, smoke, and other contaminants is the key to successful ventilation. The ultimate goal of ventilation is to remove the contaminated air and replace it with uncontaminated air. When this process is performed gradually, it is referred to as *dilution*.

Whether air is being injected (positive-pressure ventilation) or exhausted (negative-pressure ventilation) is not as important as the volume of air being moved and the displacement of the contaminated air (Figure 1.34). The volume of air being moved will affect the speed of displacement of the contaminated air. When fresh air enters a contaminated atmosphere, it mixes with the contamination, causing a dilution of the contaminants. As dilution occurs, the contaminated air is also being forced out of the area by the pressure differential. This simultaneous dilution and displacement of the contaminated atmosphere will reduce the level of contamination as rapidly as possible.

Figure 1.34 Both NPV and PPV can be used to dilute a contaminated atmosphere.

The dilution rate will depend on the following factors:

- Volume of air moved in cubic feet per minute (ft^3/min) (m^3/min)
- Proper placement of the fan (if used)
- Size of the openings being used
- Availability of replacement air

The greater the volume of air moved per minute, the faster the dilution of the atmosphere. According to the dilution principle, as air is introduced into a space, it mixes with the contaminated air present in the room, diluting it as it is being expelled (Figure 1.35). The rate of dilution will be reduced if additional contamination is being produced while ventilation is in progress. The rate can also be adversely affected in rooms that have no exit point for the contaminants or if the fan is improperly positioned.

It is important to remember that as the dilution process improves visibility within the space, it does not necessarily mean that all contaminants have been removed. Even if all visible products of combustion have been eliminated, the *invisible* products can still be present in dangerous concentrations. The only way to know whether all contaminants have been removed is by testing the atmosphere within the space. Self-contained breathing apparatus should be worn during all interior operations until the atmosphere has been determined to be clear of contaminants.

Replacement Air

During any ventilation operation, the availability of fresh intake or replacement air is important to effectively remove the smoke and other harmful contaminants from the building. Maintaining a source of replacement air is especially important when negative-pressure ventilation is being used.

With negative-pressure ventilation, effective air movement is dependent upon drawing fresh air in through openings in the structure. If these openings are too small to allow large volumes of replacement air to enter, churning is more likely, and the

Figure 1.35 The gradual dilution of an atmosphere within a building.

whole process will be much less efficient. *Churning* is the term used to describe the contaminated air being drawn back into the space with the replacement air because the ejector is improperly placed (Figure 1.36).

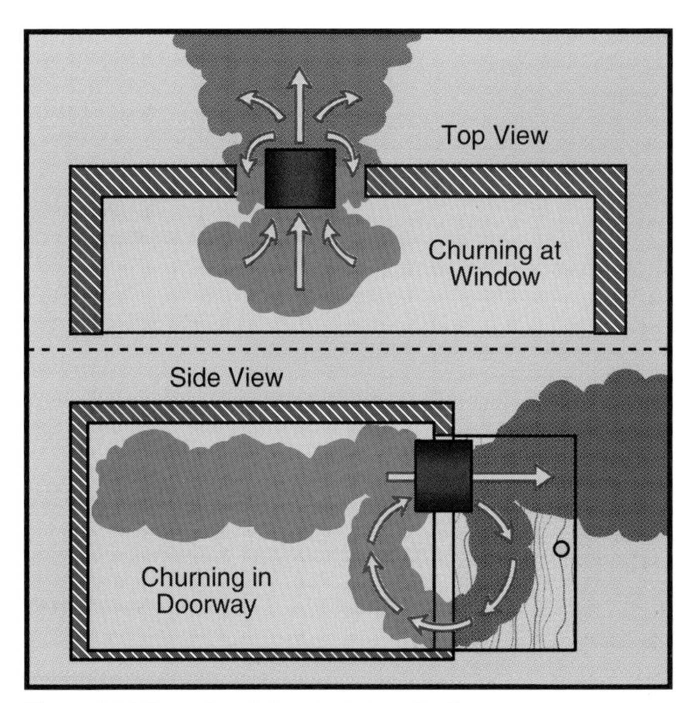

Figure 1.36 Illustration of churning in two situations.

Churning is not a problem when positive-pressure ventilation is used because the blower is taking its source of fresh air from the outside. The same concept does, however, apply when providing exit openings for exhausting smoke or contaminants to the outside. Positive-pressure ventilation will not work efficiently if fresh, replacement air is being introduced into a building with an inadequate outlet for the contaminants.

During a fire, the burning process draws in air from every opening in order to support itself, so initiating ventilation procedures may increase the availability of air to support combustion. This procedure will allow the fire to intensify, so it is very important to have a general idea of where the fire is located and to have attack lines manned and ready prior to initiating ventilation procedures.

Chapter 1 Review

Directions

The following activities are designed to help you comprehend and apply the information in Chapter 1 of **Fire Service Ventilation**, Seventh Edition. To receive the maximum learning experience from these activities, it is recommended that you use the following procedure:

1. Read the chapter, underlining or highlighting important terms, topics, and subject matter. Study the photographs and illustrations, and read the captions under each.
2. Review the list of vocabulary words to ensure that you know the chapter-related meaning of each. If you are unsure of the meaning of a vocabulary word, look the word up in the IFSTA **Orientation and Terminology** glossary or a dictionary, and then study its context in the chapter.
3. On a separate sheet of paper, complete all assigned or selected application and review activities before checking your answers.
4. After you have finished, check your answers against those on the pages referenced in parentheses.
5. Correct any incorrect answers, and review material that was answered incorrectly.

Vocabulary

Be sure that you know the chapter-related meanings of the following words:

- bank down *(17)*
- Btu *(13)*
- buoyancy *(12)*
- calorie *(13)*
- churning *(25)*
- circulation *(23)*
- condensation *(13)*
- dehydration *(13)*
- deleterious *(16)*
- devastating *(8)*
- diffusion *(23)*
- dilution *(24)*
- diminish *(11)*
- discernible *(10)*
- dissipate *(20)*
- emanate *(10)*
- gravitate *(20)*
- horizontal *(6)*
- hypoxia *(14)*
- incipient *(5)*
- inversion *(19)*
- joule *(13)*
- lateral *(6)*
- lethal *(13)*
- luminous *(12)*
- mitigate *(4)*
- oxidation *(8)*
- particulate *(13)*
- permeate *(23)*
- physiological *(13)*
- pulmonary *(15)*
- stratify *(19)*
- subside *(11)*
- synergistic *(14)*
- systemic *(15)*
- urea *(15)*
- vaporize *(8)*
- vertical *(11)*

Application Of Knowledge

1. List specific places (names of businesses, recreational and medical facilities, public buildings, industrial facilities, etc.) in your jurisdiction where each of the following nonfire or fire products is likely to be encountered. *(Local protocol)*
   - acetylene
   - acrolein
   - carbon dioxide
   - carbon monoxide
   - chlorine
   - hydrogen chloride
   - hydrogen cyanide
   - hydrogen gas
   - methane
   - nitric oxide
   - nitrogen dioxide
   - oxygen deficiency
   - phosgene
   - sulfur dioxide
2. Look up each of the above-listed gases in a haz mat handbook. Label each as lighter than or heavier than air.
3. Describe your department's SOPs regarding measuring toxic atmospheres and the use of SCBA. *(Local protocol)*

Review Activities

1. List in order and briefly describe the three phases of fire. *(5-8)*
2. Explain the ventilation procedures generally required during the first or beginning phase of a fire. *(6)*
3. Distinguish among rollover, flashover, and backdraft. *(6-8)*
4. Describe ventilation practices used to reduce the chances of rollover. *(6)*
5. Describe attack and ventilation techniques used to reduce the chances of flashover. *(7)*
6. State why proper ventilation techniques are critical during the hot-smoldering phase of a fire. *(8)*
7. Explain why opening a door or window during the late steady-state phase of a fire may cause a backdraft. *(8)*
8. List warning signs of possible backdraft that can be observed from the outside of the building. *(9)*
9. List warning signs of possible backdraft that are discernible inside the building. *(10, 11)*

10. Describe safety precautions firefighters can take if they find themselves in a building where backdraft conditions are developing. *(11)*
11. Describe ventilation practices used on buildings in which backdraft conditions have developed. *(11, 12)*
12. List and briefly explain the four major products of combustion. *(12, 13)*
13. Distinguish between the terms *temperature* and *heat.* *(13)*
14. State the Law of Heat Flow. *(13)*
15. Distinguish among a carcinogen, a toxicant, an irritant, and an asphyxiant. Provide specific examples of each. *(13, 14)*
16. Name some specific particulates that may be present in smoke. (14)
17. Explain the significance of vapor density as it relates to fire gases. *(13)*
18. Explain why the continued use of protective breathing apparatus is necessary during overhaul. *(13, 14)*
19. List the factors that determine what toxic gases will be given off at a fire. *(14)*
20. Briefly identify each of the following fire gases: *(15)*
    - acrolein
    - carbon dioxide
    - carbon monoxide
    - hydrogen chloride
    - hydrogen cyanide
    - nitric oxide
    - nitrogen dioxide
    - phosgene
21. Describe the harmful effects on the human body of each of the fire gases listed in Activity 20. *(15)*
22. Explain each of the following methods of heat transmission, and provide a specific example to illustrate each.
    - conduction *(16)*
    - convection *(16)*
    - radiation *(16, 17)*
23. Briefly identify each of the following fire behaviors:
    - mushrooming *(17)*
    - pressure differential *(18)*
    - thermal layering *(18)*
    - inversion *(19)*
24. List several nonfire gases, and explain where each is most likely to be found in the community. *(19, 20)*
25. List several examples of heavier-than-air gases and several examples of lighter-than-air gases. Explain where these gases may be found in the community. *(20)*
26. Explain the different ventilation techniques used on heavier-than-air and lighter-than-air gases. *(20, 21)*
27. Explain what firefighters must assume about airborne contaminants when working in a confined space. *(21)*
28. Briefly explain how pressure is transferred from one area to another. *(21, 22)*
29. Define and explain the term *neutral pressure plane.* *(22)*
30. Define *stack effect*, and describe what causes it. *(22)*
31. Define *diffusion*, *circulation*, *dilution*, and *churning* in relation to airflow characteristics. *(23, 24)*
32. List the four factors that affect dilution rate. *(24)*

Questions And Notes

Ventilation Size-Up

This chapter provides information that addresses the following performance objectives of NFPA 1001, *Standard for Fire Fighter Professional Qualifications* (1992):

Chapter 3 — Fire Fighter I

3-9.1 Define the principles of ventilation, and identify the advantages and effects of proper ventilation.

3-9.2 Identify the safety considerations and precautions to be taken in performing ventilation.

3-9.3 Describe the advantages and disadvantages of the following types of ventilation:
- (a) Vertical
- (b) Horizontal
- (c) Trench/strip
- (d) Mechanical
- (e) Mechanical pressurization
- (f) Hydraulic

3-9.10 Define procedures for the types of ventilation referred to in section 3-9.3.

Chapter 4 — Fire Fighter II

4-9.3 Identify considerations that must be made when determining the location and size of a ventilation opening, including:
- (a) Availability of natural openings
- (b) Location of the fire
- (c) Direction in which the fire will be drawn
- (d) Type of building construction
- (e) Wind direction
- (f) Progress of the fire
- (g) Condition of the building
- (h) Obstructions
- (i) Relative efficiency of large vs. small openings

4-9.5 Identify fire ground situations where forced ventilation procedures may be required.

Safety Points

In its discussion of ventilation size-up, this chapter addresses the following safety points:

- In assessing the life hazard in a structure fire, the incident commander (IC) must remember to consider the firefighters as well as the occupants.
- When sizing up a ventilation situation, the IC must weigh the benefits to be gained by ventilating the building against the risk that firefighters would have to take in the ventilation operation.
- The IC must weigh the effects of ventilating horizontally against the risk of placing firefighters on the roof for vertical ventilation.
- When firefighters are on the roof of a burning building, they and the IC must continually assess the roof's structural integrity.
- When firefighters are working on a roof, the IC must consider the physical effects of the weather on the vent group personnel.

Chapter 2
Ventilation Size-Up

INTRODUCTION

When evaluating ventilation situations, incident commanders (ICs) must use all of their knowledge, training, and experience relating to ventilation to arrive at the best possible solutions (Figure 2.1). While some situations may involve confined space rescues and other nonfire incidents, the vast majority of ventilation problems encountered by firefighters will involve ventilating structure fires; therefore, the main focus of this chapter is fireground operations. Typically, ventilation considerations involve basic size-up and making certain decisions:

Size-up
- Assessing rescue/life hazards
- Determining the location and extent of the fire
- Identifying building construction features

Figure 2.1 An IC assessing a ventilation situation.

Decisions
- Assessing the need for ventilation
- Deciding where ventilation is needed
- Deciding how ventilation should be accomplished

ASSESSING RESCUE/LIFE HAZARDS

The first consideration in any emergency operation is the safety of occupants and firefighters. In a structure fire, one of the best ways of minimizing hazards is to restore a tenable atmosphere to the building — especially to escape routes — through early and effective ventilation (Figure 2.2). Depending on conditions, it may be necessary to ventilate first in order to draw heat and smoke away from occupied areas, or it may be better to attack the main body of the fire immediately. Often, both must be done simultaneously. To perform these diverse functions simultaneously requires that sufficient personnel resources and equipment be on

Figure 2.2 Early and effective ventilation can restore a tenable atmosphere to a building.

the scene as soon as possible, and these should be part of the initial response.

Both occupants and activities must also be considered when assessing the rescue/life hazard. The IC needs to have reliable data about the number and locations of people normally occupying the building at the time of the fire, as well as data about the types of activities that normally go on in the building. Obviously, the life hazard in an elementary school is quite different at night or on weekends than during normal school hours (Figure 2.3). On the other hand, the life hazard in residences, hotels, and apartment buildings is significantly higher at night than during the workday.

Figure 2.3 An elementary school has a different life hazard during weekends and at night than during normal school hours.

Once rescue requirements have been assessed, the next consideration is to determine the presence of hazardous materials, processes, or conditions that may endanger fire fighting crews (Figure 2.4). Because it is impossible to accurately determine the toxicity of smoke under fire conditions, pre-incident planning should provide a familiarity with the types of materials normally involved in each major building within the response area. The presence of heavy stock or machinery on upper floors or roofs must also be considered (Figure 2.5). In addition to the visits scheduled in the pre-incident planning program, firefighters should monitor business and industrial occupancies within their district and should visit these buildings whenever their occupancy changes.

The ventilation operation must be coordinated with rescue and/or fire attack. This coordination is important because the fire often increases in intensity when ventilation is initiated because of the inrush of oxygen-rich air replacing fire gases and smoke that leave through exit openings. Charged hoselines should be in place to cover both internal exposures (uninvolved areas of the fire building) and external exposures (adjacent buildings or other combustibles) before the building is ventilated (Figure 2.6). Visibility improves as smoke and gases are expelled, so rescue and suppression crews should be ready to immediately move into the structure to complete their respective tasks.

Figure 2.4 The presence of hazardous materials and industrial processes should be determined in pre-incident planning.

Figure 2.5 Highly stacked rolls of newsprint, each weighing thousands of pounds.

Figure 2.6 Charged hoselines in place to cover both internal exposures and external exposures.

One of the most important things to remember when ventilating vertically or horizontally is: *Never* direct water streams into ventilation exit openings (Figure 2.7). Because large quantities of smoke and fire will often issue from exit openings and because firefighters are trained to extinguish fire, a natural impulse for inexperienced firefighters is to use hoselines to douse these flames. This action can be devastating to rescue and fire fighting crews inside. Directing a water stream into a building through a ventilation exit opening obstructs the natural flow of smoke, steam, and other gases out of the building and keeps them inside, defeating the purpose of ventilation. Visibility, which should be improving, suddenly degenerates as the natural convection currents are upset. The steam that is generated can burn the firefighters inside (even though they are wearing protective clothing), heat again becomes trapped within the structure, and the atmosphere quickly deteriorates from tenable to untenable.

Once ventilation has been completed, the ventilation crews should prepare to be assigned to other tasks (Figure 2.8). This reassignment will help ensure that ventilation personnel are not unnecessarily exposed to combustion products and, in the case of vertical ventilation, to the risk of roof collapse.

Figure 2.7 A hose stream should never be directed into a structure through ventilation openings.

Figure 2.8 Ventilation team being reassigned after ventilation has been completed.

DETERMINING THE LOCATION AND EXTENT OF FIRE

A fire may have spread some distance within a structure by the time fire fighting forces arrive, and consideration must be given to the extent of the fire as well as to its location. The severity and extent of the fire usually depend upon the type of fuel burning, the installed fire protection devices (or lack thereof), the degree of confinement of the fire, and the length of time it has been burning. These variables will determine the phase to which the fire has progressed, whether it be incipient, free burning or steady state, or hot-smoldering. Heavily involved free-burning fires may be beyond the point where an interior attack can reasonably be mounted, making ventilation too risky and of too little value to attempt (Figure 2.9). With few exceptions, confined fires in the hot-smoldering phase *must* be vented before any further action is attempted.

Figure 2.9 A fire far beyond the need for ventilation.

In most instances, ventilation should not be carried out until the seat of the fire has been located. Smoke that is coming out of the top floor does not always indicate a fire on the top floor. The fire may be on a lower floor or even in the basement. Creating openings for ventilation purposes before the fire is located may cause the fire to spread throughout areas of the building that otherwise would not have been affected. However, it will sometimes be necessary to ventilate in order to find the seat of the fire, such as when firefighters have searched a building filled with smoke but cannot locate the source of the smoke.

Hidden fire may be located by feeling a wall or by using infrared detection equipment to locate hot spots (Figure 2.10). Extensive horizontal or vertical ventilation may be impractical or even dangerous if the location of the fire is such that ventilation will draw the fire into uninvolved parts of the building. So, whenever possible, the location of the fire should first be accurately determined by a systematic size-up.

Figure 2.10 A firefighter using an infrared detector to locate hot spots.

Visible Smoke Conditions

Smoke accompanies most ordinary forms of combustion, and it differs greatly with the nature of the substances being burned and the amount of available oxygen. Because of these variables, firefighters must learn to recognize different types and phases of fire by observing the smoke in order to protect themselves from rollovers, flashovers, and backdrafts. This recognition is important because a free-burning fire must be treated differently than one in the hot-smoldering phase, and a fire that is relatively small in size can be mistaken for a larger fire on account of the volume of smoke (Figure 2.11). In addition to the gaseous products of combustion, other constituents, such as tar, unburned carbon, and ash, are drawn upward by the draft created by the heat of the fire. The density and color of the smoke is directly proportional to the amount of these suspended particles.

Figure 2.11 Even a small trash can fire can produce a large volume of smoke.

A fire that is just starting and is consuming wood, cloth, paper, or other ordinary combustibles will usually give off gray-white or blue-white smoke of no great density. As the burning progresses, the density may increase, and the smoke may become darker in color because it contains large quantities of carbon particles. Black smoke is usually the result of burning hydrocarbons such as rubber, roofing tar, oil, or some plastics (Figure 2.12). Brown or copper-colored smoke may indicate the presence of oxides of nitrogen.

WARNING

Gray-yellow smoke is one danger signal of a potential backdraft.

Although the color of the smoke may be of some value in determining what is burning (as well as where and to what extent), it is only a general indicator because the precise makeup of smoke can only be determined by chemical analysis. On the fireground, firefighters must assume the worst about the content of the smoke until proven otherwise.

Figure 2.12 Heavy black smoke produced by burning roofing material. *Courtesy of Bob Norman.*

When a fire is generating a lot of heat, smoke, and gases, smoke may be seen coming from attic vents or through openings around skylights, penthouses, and scuttles. Smoke may issue from under roof coverings or through small openings in the walls (Figure 2.13). It may be lazily drifting out into the atmosphere, or it may be coming out under pressure. The speed with which smoke emerges is an indication of conditions inside. The greater the pressure of the smoke exiting the building, the greater the need for rapid and effective ventilation.

Heat Conditions And Fire Severity

Smoke conditions are not always easily observed, especially at night, so other indicators of the fire's status must be used to determine its

Figure 2.13 Smoke being forced out through small cracks in the walls of a building.

location and extent. As a fire progresses, high temperatures develop. Unless the fire vents itself by breaking a window or burning through to the outside, these high temperatures continue to build well into the smoldering phase. Firefighters may obtain an indication of the intensity of the heat by feeling walls, doors, or windows; by looking for discolored or blistered paint; or by using infrared heat detectors (Figure 2.14). Hot spots give an indication of the travel of heat or the location of the fire. For example, a lower-level fire may sometimes be located by a hot spot on the floor above it, or an attic fire may be revealed by an area of melted snow on the roof. Where an involved building is one of a connected series in a business block, checking the interior walls or exterior windows of adjoining occupancies may disclose the fire's location. Knowing the exact location and severity of the fire is critical to making sound ventilation decisions.

Figure 2.14 A firefighter feeling a window for heat.

IDENTIFYING BUILDING CONSTRUCTION FEATURES

In addition to the location and extent of the fire, the building type, design, and occupancy are also factors that must be considered when making fireground ventilation decisions. Type and design features that have a bearing on these decisions include the types of materials of which the building is constructed, the number and size of both roof vents or openings and wall openings, the number of stories, and the availability of roof access (Figure 2.15). Also to be considered are security devices on window and door openings, as well as the direction in which the openings face in relation to exposures and the prevailing wind. Building construction features are discussed in greater detail in Chapters 3 and 4.

Figure 2.15 Building type and design can have a bearing on ventilation decisions.

ASSESSING THE NEED FOR VENTILATION

One of the most critical decisions that must be made in developing a plan of attack is whether or not ventilation is needed. Before the IC orders a ventilation operation, he or she must consider the effects that ventilation will have on the intensity of the fire. Prior to actually opening the building, consideration should be given to the state of readiness of on-scene personnel to advance hoselines for facilitating rescues, for fire attack, and for exposure protection.

As part of determining the overall strategy for a particular incident, the IC must determine whether ventilation is necessary and if so, when and in what form it should be initiated. Ventilation is a tactical operation, and its purpose is to direct the movement of the products of combustion and to expel them from the building to facilitate rescue and fire attack. Some situations may simply require locating and extinguishing the fire and then ventilating afterward to clear residual smoke from the structure. Others will require early and aggressive ventilation to allow search/rescue and fire suppression personnel the opportunity to safely work inside the building (Figure 2.16).

Figure 2.16 Firefighters quickly setting up positive pressure ventilation.

DECIDING WHERE VENTILATION IS NEEDED

The point at which smoke emerges from a building may offer some indication of where to ventilate. Smoke coming from a lower story may indicate an absence of vertical openings above the point where the smoke appears. Smoke coming through only one end of the building may indicate the location of the fire or may be the result of wind influence.

Regardless of what location is selected for the ventilation openings, the first consideration should be the safety of firefighters and occupants. The exit opening should be as close to the seat of the fire as possible. If vertical ventilation is selected, the roof should be opened directly over the center of the fire (Figure 2.17). If this position is unsafe, the hole should be made between the fire and the uninvolved portion of the building in a location where firefighters will not be placed in jeopardy. If neither of these options is possible, a defensive tactic (such as trench/strip ventilation — see Chapter 4) may be considered (Figure 2.18). Some of the factors that have a bearing on where to ventilate include the following:

- Location of the fire
- Location of occupants
- Interior and exterior exposures
- Type of construction
- Purpose or use of occupancy
- Extent to which the fire has progressed

Figure 2.17 Firefighters attempting to ventilate directly over the seat of the fire.

Figure 2.18 Trench ventilation may be necessary to cut off the lateral spread of a fire. *Courtesy of Douglas F. Shelby.*

- Condition of the building and its contents
- Existing openings (skylights, ventilator shafts, monitors, etc.)
- Direction of the wind
- Available personnel and equipment

The IC must consider each of these factors and decide what would be the most effective location for ventilation in terms of protecting occupants and firefighters, protecting exposures, and facilitating fire suppression.

DECIDING HOW VENTILATION SHOULD BE ACCOMPLISHED

This decision will be the result of considering where to ventilate (based on the demands of the situation) and a knowledge of the capabilities and limitations of various ventilation methods. The major considerations are whether to ventilate horizontally or vertically and whether to use natural or forced ventilation. While some aspects of vertical ventilation may also apply to horizontal ventilation, ventilating a room, a floor, a cockloft, an attic, or a basement will each require somewhat different procedures. A more complete discussion of horizontal ventilation is presented in Chapter 3, vertical ventilation in Chapter 4, and natural versus forced (mechanical) ventilation in Chapter 5.

Vertical Versus Horizontal Ventilation

Horizontal ventilation is accomplished by opening windows or doors to allow smoke to escape and fresh air to enter (Figure 2.19). Structures in which horizontal ventilation may be appropriate include the following:

- Buildings in which the fire is not large enough to necessitate opening of the roof
- Buildings with windows or doors close to the seat of the fire
- Buildings in which the fire involvement is below the top floor
- Buildings in which fire has not entered structural voids or concealed spaces

Vertical ventilation is accomplished by opening a structure at the highest point by cutting holes in roofs or by opening doors, scuttles, or skylights to allow the heat and smoke to travel upward and out of the structure (Figure 2.20). Structures in which vertical ventilation may be appropriate include the following:

- Buildings with fire in the attic, the cockloft, or the top floor
- Windowless buildings with few exterior doors
- Buildings with large vertical shafts (light wells, elevator hoistways, etc.)
- Buildings in which fire has entered structural voids or concealed spaces

Natural Versus Forced Ventilation

Ventilation efforts should be in concert with existing atmospheric conditions, taking advantage of natural ventilation whenever possible. If condi-

Figure 2.19 Horizontal ventilation through a window opening.

Figure 2.20 Firefighters working on a roof to ventilate vertically. *Courtesy of Edward Prendergast.*

tions are favorable, natural ventilation is fast and efficient because it requires no additional personnel or equipment to set up and maintain (Figure 2.21). However, in some situations natural ventilation may be inadequate and may have to be supplemented or replaced by forced ventilation to facilitate rescue and suppression operations and to provide a tenable atmosphere. Forced or mechanical ventilation involves the use of fans, blowers, nozzles, or other mechanical devices to create or redirect the flow of air inside an involved space.

Figure 2.21 A firefighter taking out a window for natural horizontal ventilation.

Using forced ventilation eliminates or reduces the effect of unstable and erratic winds on ventilation efforts within a structure. With a dependable, controllable airflow provided, greater control of the movement of heat and smoke is possible. Forced ventilation can channel the airborne products of combustion or other contaminants out of a building by the most efficient and least destructive path and allow fresh air to be reintroduced into the space (Figures 2.22 and 2.23). In both fire and nonfire situations, using proper forced ventilation in conjunction with natural ventilation allows a tenable atmosphere to be restored faster and more efficiently than with natural ventilation alone.

SITUATIONS REQUIRING FORCED VENTILATION

Whenever natural ventilation is inadequate or inappropriate or if other elements in the situation

Figure 2.22 A smoke ejector being used to exhaust contaminants from a building.

Figure 2.23 A blower being used to reintroduce fresh air into a space while forcing smoke out.

limit or preclude natural ventilation, forced ventilation should be considered. Otherwise, there are no definite rules governing when it should or should not be employed. In general, forced ventilation is indicated in the following situations:

- When the location and extent of fire has been determined
- When the type of construction is not conducive to natural ventilation
- When natural ventilation slows, becomes ineffective, and needs support
- When fire is burning below grade in structures or below deck in marine vessels
- When a contaminated atmosphere in nonfire situations must be cleared from a confined space
- When the contaminated area within a confined space is so large that natural ventilation is impractical or too inefficient

Figure 2.24 Occupants may have to use the same routes that smoke and heated gases travel to reach exterior openings.

OTHER CONSIDERATIONS

Other items that need to be considered in sizing up a ventilation situation are exposures, both internal and external, and the weather. Ventilation operations can be seriously affected by wind, humidity, and temperature.

Exposures

Because horizontal ventilation does not normally release heat and smoke directly above the fire, some routing of the smoke and fire gases to an exterior opening may be necessary. The IC must consider the threat to internal exposures that this routing can create. The routes by which the smoke and heated gases travel to reach the exterior openings may be the same corridors and passageways that occupants must use for evacuation (Figure 2.24). In horizontal ventilation, fire and heated gases are released through window openings or doorways. Consequently, there is the constant danger that they will ignite the structure above the point where they escape or that they may be drawn into windows above the ventilation opening (lapping or autoexposure) (Figure 2.25).

Figure 2.25 A fire may lap into uninvolved areas above the fire.

The IC must also consider the possible threat to external exposures from ventilation efforts. Horizontal ventilation threatens exposed buildings through radiation and/or direct flame contact (Figure 2.26). Smoke may also be drawn into adjacent buildings by their air-conditioning units. Vertical ventilation may also threaten nearby structures if hot brands or embers that are carried aloft by convection fall onto combustible roofs or into dry vegetation, or if adjacent structures are taller than the fire building (Figure 2.27).

Figure 2.26 Ventilating horizontally may spread the fire by radiated heat and/or direct flame contact.

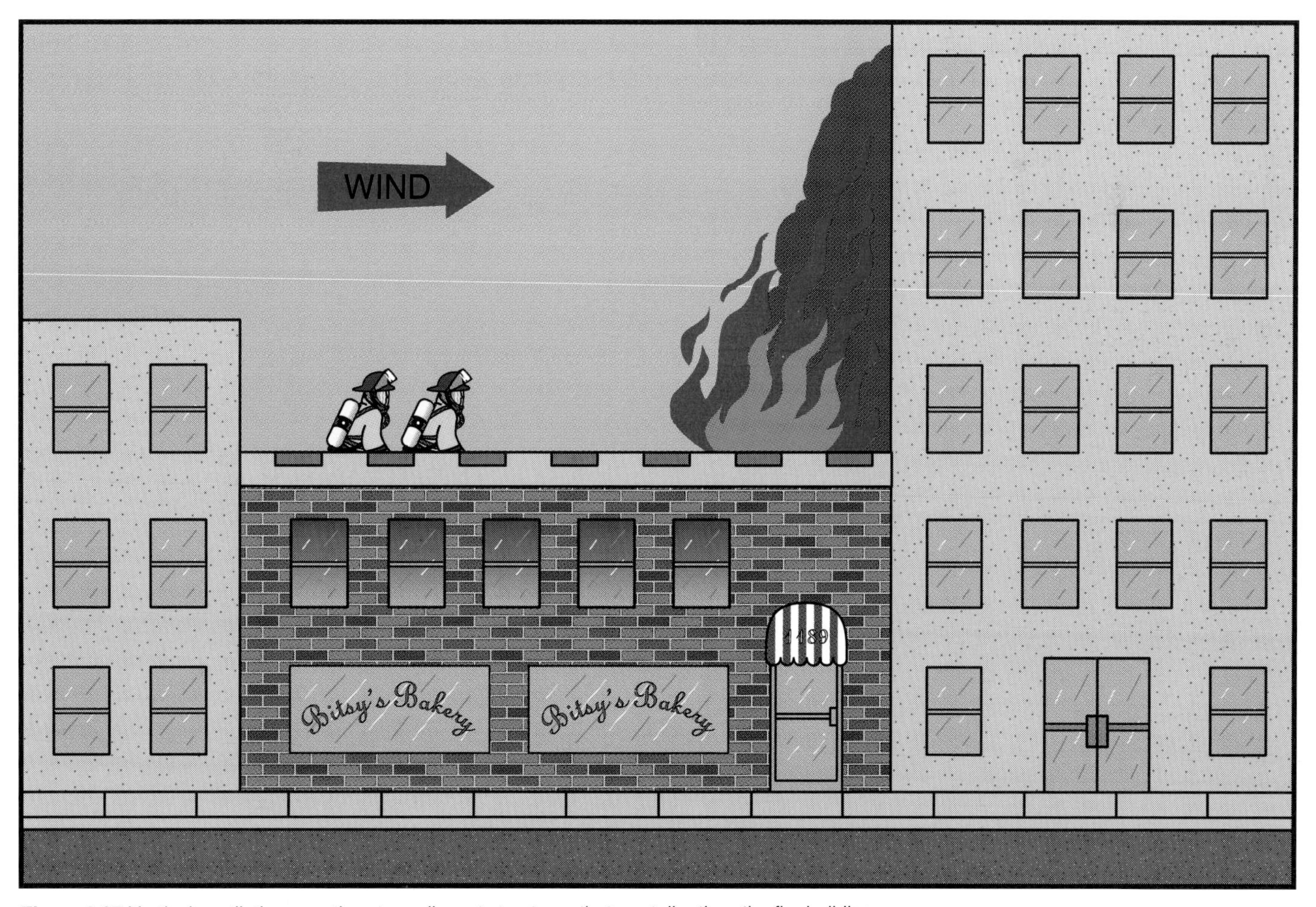

Figure 2.27 Vertical ventilation may threaten adjacent structures that are taller than the fire building.

Weather Conditions

WIND

Wind conditions are always a factor in determining the proper ventilation procedure. Wind direction may be designated as windward or leeward. The side of the building which the wind strikes is *windward*; the opposite side is *leeward* (Figure 2.28). Because of the dangers of wind blowing fire toward an external exposure, feeding oxygen to the fire, or blowing the fire into uninvolved areas of the building, problems may be encountered when using horizontal ventilation.

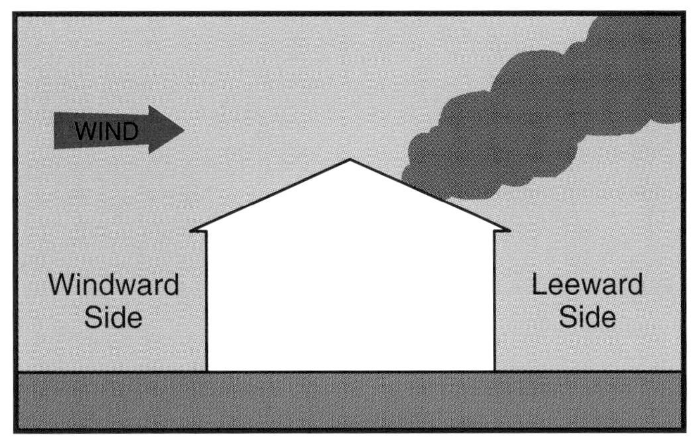

Figure 2.28 Illustration of windward side and leeward side of a building.

HUMIDITY

High humidity tends to keep the products of combustion from rising into the atmosphere, thus keeping these products at or near ground level (Figure 2.29). When high humidity creates these conditions, ventilation will be more difficult, will be more dangerous to firefighters, will be more time consuming, and will almost certainly require the use of forced ventilation.

TEMPERATURE

While the atmospheric temperature has little direct effect on the behavior of fire within a structure, on a hot summer day it can have a profound effect on the ventilation operation. The temperature on the flat roof of a commercial building — often black or dark gray and surrounded by a parapet wall — can be significantly higher than the temperature at street level. The effects of the elevated temperature on ventilation personnel, in full turnouts and doing very strenuous work, must be considered. The crews are at risk of heat exhaustion or even heatstroke, and their productivity will be reduced because of fatigue (Figure 2.30). This fatigue can delay the completion of their assign-

Figure 2.29 Night inversion keeping smoke from rising.

ments and will require earlier and more frequent rest breaks and/or crew reliefs.

Cold atmospheric temperatures and the conditions they produce may also affect ventilation operations. In winter, snow and ice accumulations on the roofs of buildings can increase live loads, conceal roof features, hamper access, and delay the completion of vertical ventilation operations. The dangers to firefighters may also be increased due to poor footing on wet or icy roofs.

Figure 2.30 Firefighters cooling off in Rehab. *Courtesy of Ron Jeffers.*

Chapter 2 Review

Directions

The following activities are designed to help you comprehend and apply the information in Chapter 2 of **Fire Service Ventilation**, Seventh Edition. To receive the maximum learning experience from these activities, it is recommended that you use the following procedure:

1. Read the chapter, underlining or highlighting important terms, topics, and subject matter. Study the photographs and illustrations, and read the captions under each.
2. Review the list of vocabulary words to ensure that you know the chapter-related meaning of each. If you are unsure of the meaning of a vocabulary word, look the word up in the IFSTA **Orientation and Terminology** glossary or a dictionary, and then study its context in the chapter.
3. On a separate sheet of paper, complete all assigned or selected application and review activities before checking your answers.
4. After you have finished, check your answers against those on the pages referenced in parentheses.
5. Correct any incorrect answers, and review material that was answered incorrectly.

Vocabulary

Be sure that you know the chapter-related meanings of the following words:

- adjoining *(36)*
- aggressive *(37)*
- conjunction *(39)*
- defensive *(37)*
- erratic *(39)*
- expel *(32)*
- facilitate *(37)*
- jeopardy *(37)*
- occupancy *(32)*
- prevailing *(36)*
- residual *(37)*
- tactical *(37)*
- tenable (untenable) *(31)*

Application Of Knowledge

1. Choose a public building in your jurisdiction. Walk through the building. Observe and take notes on the occupants, activities, furnishings, and construction features of the building.

   Assume that fire breaks out on the southwest corner of the building's ground floor at 3 p.m. on a calm, sunny weekday afternoon. Based on your observations, describe probable rescue/life safety hazards that you would be likely to encounter. Discuss ventilation procedures you might initiate. *(Local protocol)*

2. Study the pre-incident survey for a building in your jurisdiction. Identify building construction features and external exposures that would affect ventilation methods or location. Decide where, how, and what type of ventilation you might use were there a fire in the northeast corner of the first floor of the building at 6 p.m. on a rainy, windy fall weekend. *(Local protocol)*

Review Activities

1. Briefly identify each of the following:
   - size-up *(31)*
   - infrared detection equipment *(34)*
   - internal exposures *(32)*
   - external exposures *(32)*
   - hydrocarbon *(35)*
   - oxides of nitrogen *(35)*
   - cockloft *(38)*
   - scuttle *(35)*
   - hot spot *(34)*
   - lapping *(40)*
   - autoexposure *(40)*
   - brand *(42)*
   - radiation *(40)*
   - convection *(33 & 40)*
2. Distinguish between offensive and defensive ventilation tactics. Provide examples of each. *(37)*
3. List guidelines for assessing rescue/life safety hazards. *(31-33)*
4. Explain why the ventilation operation must be coordinated with rescue and fire attack. *(32)*
5. Describe what may happen if an inexperienced firefighter directs a water stream into a ventilation opening. *(33)*
6. List factors upon which the severity and extent of a fire usually depend. *(34)*
7. Explain why ventilation is not carried out, in most instances, until the seat of the fire is located. *(34)*
8. Describe what a firefighter can learn about a fire's status by observing the color, density, and pressure of smoke exiting a building. *(34, 35)*
9. List ways that firefighters determine the intensity of a fire or its location. *(36)*

10. List building type and design features that affect ventilation decisions. *(36)*
11. Explain how a firefighter determines whether or not to ventilate. *(36, 37)*
12. List factors that have a bearing on where to ventilate. *(37)*
13. Compare and contrast vertical versus horizontal ventilation. Support your discussion with specific examples. *(38)*
14. List features of structures in which horizontal ventilation may be appropriate. *(38)*
15. List features of structures in which vertical ventilation may be appropriate. *(38)*
16. Compare and contrast natural versus forced ventilation techniques. Support your discussion with specific examples. *(38, 39)*
17. List instances in which forced ventilation is appropriate. *(40)*
18. Explain how horizontal ventilation may pose a threat to both internal and external exposures. *(40, 41)*
19. Explain how vertical ventilation may pose a threat to external exposures. *(40, 41)*
20. Distinguish between the directional terms *windward* and *leeward* and between *horizontal* and *vertical. (42)*
21. Explain how wind, humidity, and atmospheric temperature may affect a ventilation operation. *(42, 43)*
22. Distinguish between heat exhaustion and heat stroke. *(42)*

    NOTE: You may have to consult the glossary of terms in IFSTA's **Orientation and Terminology** manual, an EMT text, or a medical dictionary.

Questions And Notes

3

Horizontal Ventilation

This chapter provides information that addresses the following performance objectives of NFPA 1001, *Standard for Fire Fighter Professional Qualifications* (1992):

Chapter 3 — Fire Fighter I

3-9.3 Describe the advantages and disadvantages of the following types of ventilation:
(b) Horizontal
(d) Mechanical
(e) Mechanical pressurization
(f) Hydraulic

3-9.6 Identify the types of tools used during ventilation.

3-9.10 Define procedures for the types of ventilation referred to in section 3-9.3.

3-9.11 Demonstrate opening various types of windows from inside and outside, with and without the use of tools.

3-9.12 Demonstrate breaking window or door glass and removing obstructions.

Chapter 4 — Fire Fighter II

4-9.5 Identify fire ground situations where forced ventilation procedures may be required.

Safety Points

In its discussion of horizontal ventilation, this chapter addresses the following safety points:

- Firefighters must understand the capabilities and limitations of the various ventilation tools and be well trained in their safe operation.
- To avoid initiating a backdraft when attempting to ventilate horizontally, firefighters must be able to recognize the signs of backdraft conditions.
- To avoid injury from broken glass, firefighters must know and use proper techniques when breaking windows.
- To avoid upsetting established horizontal ventilation, firefighters must understand the importance of ensuring that doors being used as ventilation openings cannot be closed inadvertently and the importance of not directing hose streams into ventilation exit openings.
- To avoid respiratory injury, firefighters must wear SCBA when performing hydraulic ventilation within a structure.

Chapter 3
Horizontal Ventilation

INTRODUCTION

Horizontal ventilation is the most frequently used form of ventilation because in the majority of situations it is the most appropriate method with which to ventilate the building. Also, firefighters entering a building for search and rescue or fire attack start a form of horizontal ventilation by opening doors or windows to make entry. Statistically, most fires in buildings are small, nuisance fires that do little damage and produce more smoke than fire. These situations may only require that the windows and doors of the affected occupancy be opened to allow the residual smoke to be ventilated. Other fire situations, such as a working fire below the top floor of a building, may also lend themselves to horizontal ventilation.

This chapter reviews the tools and equipment needed to initiate horizontal ventilation, describes the construction of walls and horizontal openings, emphasizes the importance of establishing and supporting horizontal ventilation, and discusses the techniques involved in natural and forced horizontal ventilation.

HORIZONTAL VENTILATION TOOLS AND EQUIPMENT

Under the right conditions, practically every forcible entry tool can be applied to horizontal ventilation. In the removal or opening of doors, windows, or other barriers, the axe and more specialized prying and cutting tools may be required. Wire and bolt cutters may be needed for removing locks and security bars. Pike poles and other tools may be used for breaking windows. Openings in metal walls or roll-up doors may be made with power saws or with oxyacetylene cutting equipment. The care and use of these tools and the techniques of forcible entry are fully discussed in the IFSTA **Forcible Entry** and **Essentials of Fire Fighting** manuals.

Ventilation Tools

A wide array of hand tools and power tools can be used during horizontal ventilation operations. These range from the simple pick-head axe to the chain saw. Some tools may be used in combination to accomplish specific tasks. For example, the flat-head axe and the Halligan tool may be used together to open exterior doors (Figure 3.1).

It is important to know the limitations of ventilation tools as well as how to use and maintain them for maximum safety and efficiency. Using a tool for something other than its intended purpose can needlessly endanger firefighters and damage the tool. Regular preventive maintenance of ventilation tools improves their performance and maximizes their safety. Keeping axes and power-saw blades sharp, free of paint, and lightly oiled minimizes the effort needed to achieve desired results (Figure 3.2).

Fans

The devices necessary to supplement or alter the natural airflow in a structure or other confined space can take several different forms. Some of these items, such as smoke ejectors and blowers, are very familiar to most firefighters. Others, HVAC systems for example, may be less familiar and will require further investigation and study. For purposes of clarity, in this manual the term *ejector* will refer to any device positioned within the space or in the exit opening to blow contaminated air *out* of the space. The term *blower* will refer to any device

Figure 3.1 The tools commonly used for forcible entry can also be used for ventilation.

Figure 3.2 Cutting tools must be kept sharp to achieve the desired results.

positioned outside of a space to blow fresh air *in*. The term *fan* will be used interchangeably with blower and ejector.

Technology has created many new developments in the types of ventilation equipment available today. Fans can be driven by electric motors, gasoline-powered engines, or even water pressure (Figures 3.3a-c). These developments have led to greater efficiency and versatility, enabling firefighters to accomplish ventilation operations more rapidly than ever before. In addition, accessories, such as flexible duct attachments, stacking and hanging accessories, and other support systems, allow for flexibility in the placement of smoke ejectors.

The development of blowers has added another major option to ventilation equipment and techniques. Set up just outside the building to create a slight positive pressure within, these units can make a significant contribution to a fire department's horizontal ventilation capabilities. For a more complete discussion of different types of fans, see Chapter 5, Forced Ventilation.

Figure 3.3a Electric fan. *Courtesy of Super Vac.*

Figure 3.3b Gasoline-powered fan. *Courtesy of Tempest Technology Corporation.*

Figure 3.3c Water-powered fan. *Courtesy of RAMFAN Corporation.*

Flexible Duct Attachments

Because of the amount of time involved in setting up flexible duct ventilation systems, it is usually not an appropriate technique as part of an initial fire attack. However, using these devices to move cold smoke and/or other airborne contaminants to the outside can be very effective. This procedure allows contaminants to be channeled through a building without contaminating it. For example, with a fan and flexible duct combination, a temporary exhaust line can channel smoke down a hall, through a room, or through an entire building without causing smoke damage or contamination (Figure 3.4). Being able to channel smoke or fumes through a flexible duct to the outside without contaminating other areas is particularly useful in shopping malls where there are many unrelated occupancies and in hospitals, schools, or office buildings.

Using flexible ducting, two or more smoke ejectors can be coupled together to ventilate smoke from basements, attics, suspended ceilings, ship holds, storage bins, railroad cars, and other confined spaces. Also, ducting can provide fresh air for rescue crews involved in long-term confined space operations such as in manholes, sewers, silos, and other hostile environments (Figure 3.5).

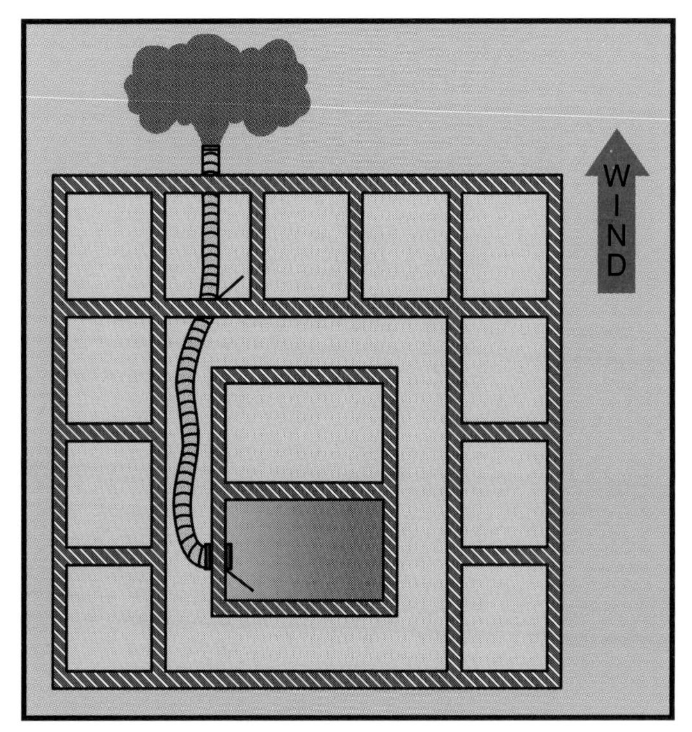

Figure 3.4 Smoke being channeled through a building by using flexible ducting.

Figure 3.5 Ducting can provide fresh air for rescue crews in confined space operations such as in manholes.

When using negative-pressure ventilation, it is important to remember that replacement air must be brought into the area from which contaminants are being exhausted. This task can be difficult when ventilating areas that have only one opening for exhaust and replacement air. However, by providing the replacement air (positively) via a flexible duct to the most distant part of the room, it is possible to exhaust the smoke and bring replacement air through the same doorway (Figure 3.6).

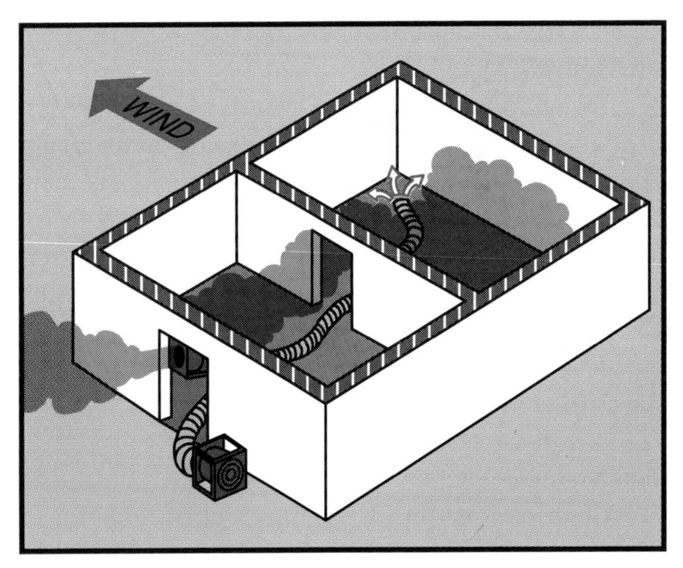

Figure 3.6 Providing replacement air by a flexible duct to the most distant part of a building.

Using the ejector/flexible duct combination is also an excellent method of ventilating areas below grade. It works well for removing heavier-than-air gases that settle near the floor and in low areas (Figure 3.7). The smoke ejector can be positioned at ground level or above, with the flexible duct running through a window or down a stairway or

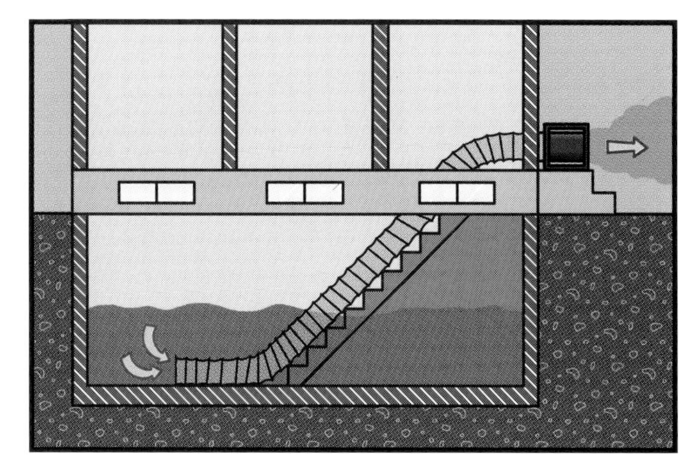

Figure 3.7 The ejector/flexible duct combination removing heavier-than-air gases from a basement.

elevator shaft into the basement. Replacement air can be channeled through the same opening in which the fan is located or through other available openings. Because of the hazards associated with heavier-than-air gases and/or oxygen deficiency, it is important to remember that even with replacement air being supplied, operations where personnel are working below grade should be constantly monitored (Figure 3.8).

Figure 3.8 A monitor double-checks all safety gear.

CONSTRUCTION RELATED TO HORIZONTAL VENTILATION

A working knowledge of standard construction practices allows firefighters to create ventilation openings for horizontal ventilation efficiently. The construction practices described in this chapter are those in general use across the country but may not be typical of those used in some local areas. Firefighters should investigate and become familiar with the construction practices used in their particular area.

Walls

Most modern buildings have continuous foundations of concrete, although older structures may have brick or stone foundations. The stem wall, located between the foundation and the first floor, may extend well above the ground if the building occupies a sloping site (Figure 3.9). Most exterior walls will have an interior finish, usually of gypsum board (commonly called *drywall*) applied over wooden or metal vertical members called *studs* (Figure 3.10). The studs are commonly spaced on 16- or 24-inch (400 mm or 600 mm) centers. Be-

Figure 3.9 A typical stem wall of an older dwelling.

tween the studs and the interior and exterior finishes are concealed spaces through which fire sometimes travels to upper floors or attics. To prevent this vertical spread of fire, these spaces sometimes contain horizontal fire-stops between the studs (Figure 3.11). These concealed spaces also often contain electrical wiring, metal or plastic plumbing, and insulation material.

Figure 3.10 A typical stud wall before the interior finish is applied.

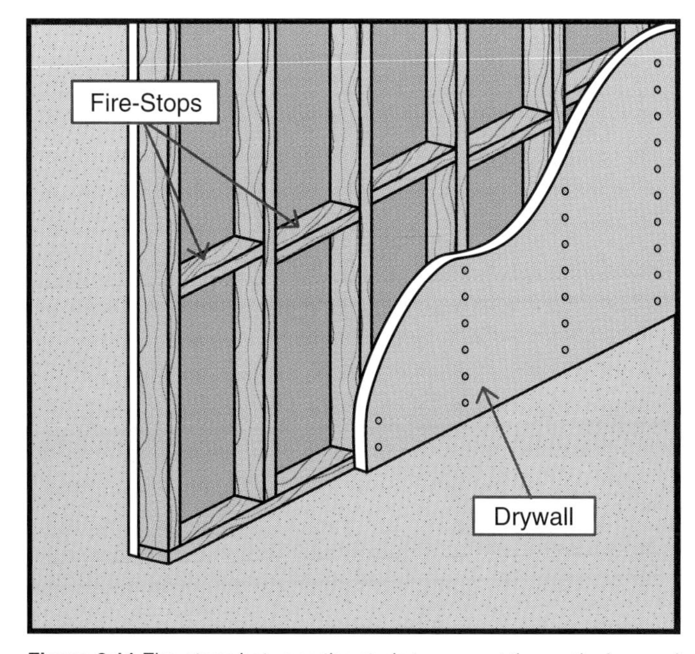

Figure 3.11 Fire-stops between the studs to prevent the vertical spread of fire.

FIRE WALLS

Fire walls are rated assemblies designed to reduce the likelihood of horizontal fire spread by compartmentalizing the building with fire-resistive separation walls. Fire walls are constructed of masonry or of a specified thickness of drywall over a wood or metal frame. They must extend the entire width of the building and, in most cases, up to and above combustible roofs (Figure 3.12). If these walls have been compromised with unprotected openings or poke-throughs for wiring or plumbing, fire may pass through these openings, requiring ventilation to be performed on both sides of the wall.

The potential for flammable fire gases to accumulate on the uninvolved side of a fire wall in an attic or cockloft emphasizes the need for early and effective ventilation of these areas. Otherwise, these gases may accumulate, unknown to the firefighter, until they reach a source of ignition. Depending on

Figure 3.12 A fire wall completely separates sections of a building. *Courtesy of Edward Prendergast.*

the availability of oxygen within the space, the ignition of these gases may result in a fire with explosive force.

Generally, fire walls should not be breached for ventilation purposes. Holes in fire walls can provide additional oxygen to the fire and/or provide a path of travel for heat, smoke, and fire gases. Drafts produced by the holes may also draw the fire toward the opening because of the availability of oxygen at the opening and on the other side of the wall.

EXTERIOR WALLS

To support the roof structure and to protect the building from the elements, most buildings have substantial exterior walls. Exterior walls may be masonry, masonry or other veneer over frame, or metal.

Masonry. Exterior masonry walls usually range from 8 to 12 inches (200 mm to 300 mm) thick, depending upon the material used. The walls may be of reinforced concrete (poured in place, tilt-up, or precast panel), concrete block, brick, or stone (Figure 3.13). Windowless masonry walls, regardless of type of material, are formidable barriers that are so difficult and time consuming to penetrate that they are rarely breached for ventilation purposes. If they must be breached, for whatever purpose, heavy-duty power equipment such as electric or pneumatic jackhammers should be used.

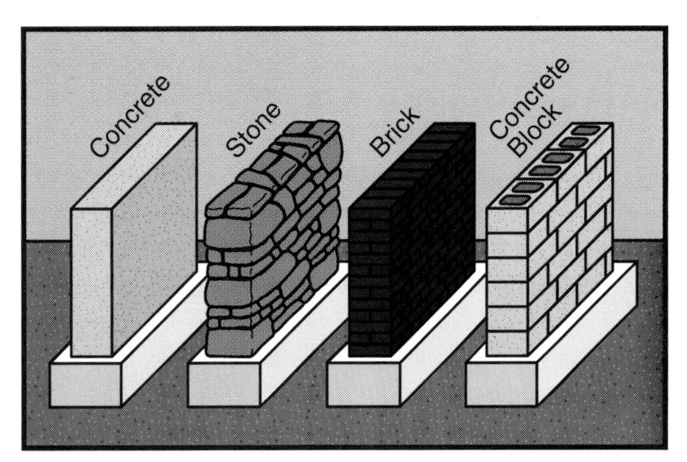

Figure 3.13 Masonry walls are constructed of concrete, stone, brick, or concrete blocks.

Veneer over frame. These walls are essentially frame walls in which the 2- x 4-inch or 2- x 6-inch structural members (studs) are wood or metal with a layer of plywood or chipboard for shear strength. One layer of brick or stone (real or imitation) is added to the exterior and/or interior surface to give the appearance of a solid brick or stone wall (Figure 3.14). A veneer of stucco may be applied over a gypsum board base or over chicken wire and tar paper directly over the studs (Figure 3.15).

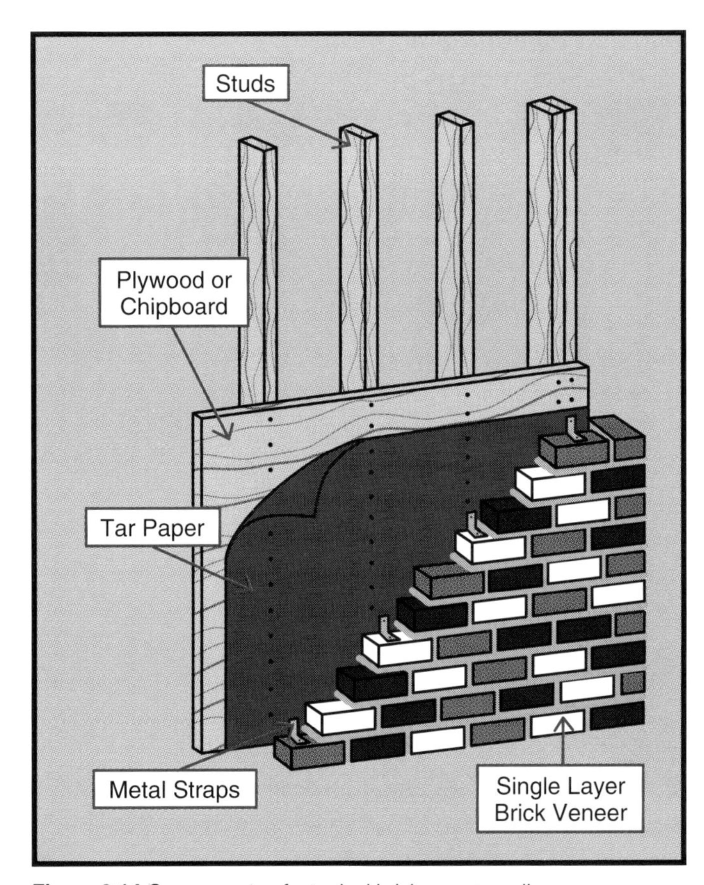

Figure 3.14 Components of a typical brick veneer wall.

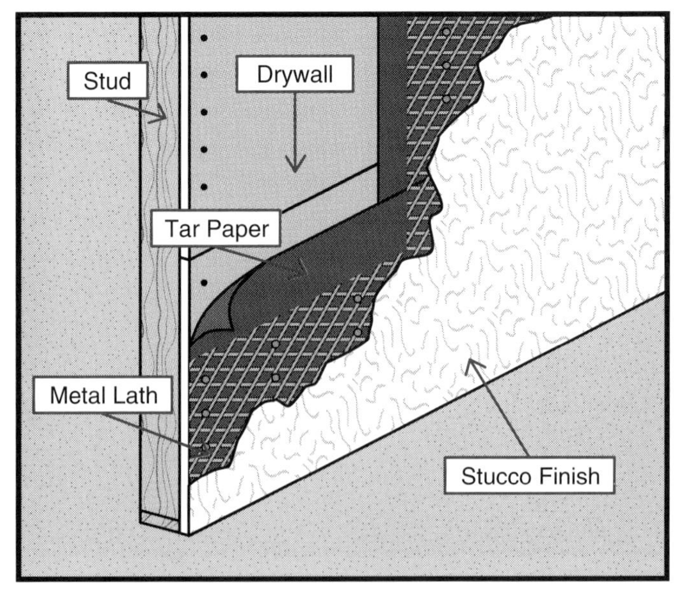

Figure 3.15 Components of a typical stucco wall.

Builders in some areas now prefer to use exterior plywood siding nailed directly to the studs because this practice provides the required shear strength and the exterior finish in one layer (Figure 3.16). In recent years, the practice of applying a veneer of Styrofoam® plastic foam to the outside of new and existing commercial buildings for aesthetic purposes has grown in popularity (Figure 3.17). This foam veneer may be blocks or sheets and range in size from a few inches (mm) to more than a foot (0.3 m) thick. After the foam is applied to the wall, it is sealed against the weather with a granular finish, giving the appearance of a solid masonry wall. These walls can be recognized by the hollow sound they make when tapped with a tool.

Unlike solid masonry walls, veneer over frame walls are much easier to penetrate with conventional forcible entry tools. Many of the same techniques used to strip roof coverings from a roof can be applied to the veneer covering of exterior walls. For example, if a windowless building with a stucco or plywood veneer must be breached for horizontal ventilation, the wall can be sounded with a tool to locate the studs, the perimeter of the opening cut with an axe or power saw, and the veneer stripped away with axes or rubbish hooks. Likewise, a circular saw equipped with a masonry blade can quickly cut through a brick or imitation stone veneer to allow the veneer to be stripped away (Figure 3.18). Masonry veneer can also be breached with sledgehammers or battering rams (Figure 3.19).

Figure 3.16 Typical exterior plywood siding.

Figure 3.17 The columns and cornices of this building are made entirely of plastic foam. *Courtesy of Wes Kitchel.*

Figure 3.18 A brick veneer wall being cut with a circular saw equipped with a masonry blade.

Figure 3.19 Firefighters about to breach a masonry veneer wall with a battering ram.

Masonry veneer walls can present a substantial collapse potential because of the way some of them are attached to the studs. Some veneers, such as brick or stone (real or imitation), are attached to the studs with thin sheet-metal straps (Figure 3.20). On a concrete block wall, one end of the strip is embedded in the mortar joint between the blocks and the other end in the mortar joints between the bricks or stones. On a wood frame wall, one end of each strap is nailed or screwed to a stud, and the other end is embedded in the mortar joint between the bricks or stones. Under fire conditions, these straps can fail because they pull out of the stud or because they snap when weakened by the heat. Firefighters should keep this potential collapse hazard in mind during breaching operations.

Figure 3.20 Sheet-metal straps nailed to the studs tie masonry veneers to the wall.

Metal. The rising costs of conventional building construction have made metal-covered buildings economically attractive, and they are now used in a wide variety of occupancies from storage to mercantile to industrial (Figure 3.21a). The exterior walls of metal buildings are usually of light-gauge sheet metal formed into rectangular panels that are applied vertically over a wood or metal frame. They may be nailed to wooden members or attached to a metal frame with screws. Some metal walls are merely curtain walls attached to a rigid structural steel frame. In this application, there usually are no studs in the exterior wall, only lightweight horizontal members between the heavy steel pillars (Figure 3.21b). These walls are easily cut with conventional hand or power tools and lend themselves to the same triangular "tepee" cut (Figure 3.22) as used for some metal roll-up doors.

Figure 3.21a A typical metal-covered building.

Figure 3.21b Horizontal frame members on the inside of the exterior walls of metal buildings.

Figure 3.22 Typical "tepee" cut in a metal wall.

Windows

For horizontal ventilation purposes, windows are some of the best means of creating the necessary exterior openings. Even when the windows cannot be opened, the panes may be broken out relatively easily and quickly. The cost of replacing the broken glass (especially if the frame is undamaged) is usually less than the cost of repairing damaged doors or walls. A wide variety of windows exists in both new and old buildings in most communities. Descriptions of some of the more common types follow.

- *Fixed windows.* As the name implies, a fixed window does not open but has a permanently glazed pane(s) set in a wooden, vinyl, or metal frame. They range from small, sometimes irregularly shaped windows to large picture windows (Figure 3.23). They are often flanked by double-hung or casement windows or are stacked with awning or hopper windows that can be opened.
- *Single- and double-hung windows.* Single-hung windows have one movable section; in double-hung windows, both halves are movable (Figure 3.24). In both cases, only half of the window area is available for ventilation unless the panes are broken out.
- *Casement windows.* These windows have one or two side-hinged, outward-swinging sashes, and the screens are on the inside (Figure 3.25). Double casement windows may be separated by a fixed pane or simply by a vertical post called a *mullion*. The entire window area is available for ventilation when open.
- *Horizontal sliding windows.* These windows have two or more sashes, one of which is fixed and the other(s) movable (Figure 3.26). From the inside, the movable sash can often be lifted out of the frame without damaging the window. In most designs, half of the window area is available for ventilation without breaking the panes.

Figure 3.25 A typical casement window with two side-hinged, outward-swinging sashes.

Figure 3.23 A typical metal frame window with a fixed middle section. The sections on each end slide horizontally.

Figure 3.24 A typical double-hung window. Both sections are movable vertically.

Figure 3.26 A typical horizontal-sliding window with one fixed sash and one movable sash.

- *Awning windows.* These windows have one or more top-hinged, outward-swinging sashes (Figure 3.27). Single awning windows are often combined with a fixed sash in a larger unit. All of the openable area is available for ventilation.
- *Jalousie windows.* These windows consist of narrow horizontal panes of glass set in pivoting brackets at each end (Figure 3.28). The panes overlap in a shinglelike fashion. They are very difficult to open from the outside without breaking the panes. When open, they offer the entire area for ventilation.
- *Projected windows.* These windows, also known as *factory windows*, may be hinged at the top or bottom and may swing inward or outward (Figure 3.29). All of the openable area is available for ventilation.

Figure 3.27 A typical awning window with three top-hinged, outward-swinging sashes.

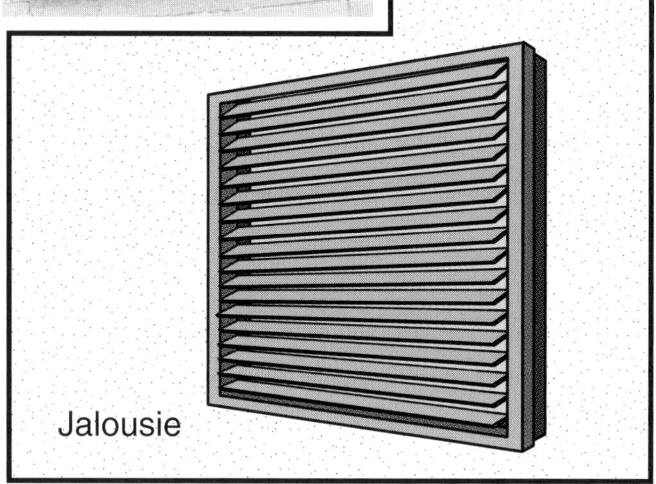

Figure 3.28 A typical jalousie window with narrow horizontal panes of glass set in pivoting brackets at each end.

- *Hopper windows.* These windows have bottom-hinged, inward-swinging sashes (Figure 3.30). They are often just awning windows that have been installed upside down. Like awning windows, they offer all of the openable area for ventilation.

Figure 3.29 A typical projected (factory) window hinged at the top and swinging outward.

Figure 3.30 A typical bottom-hinged hopper window.

Some *energy-efficient windows* are double- or triple-glazed and have an airspace (or, sometimes, a vacuum) between the panes (Figure 3.31). Although Thermopane® is one brand of energy-efficient windows, the name is commonly applied to all such windows. While these windows offer some additional protection from an exposure fire, they also present some disadvantages for firefighters.

Because of their superior insulating properties, energy-efficient windows hold in more heat than conventional windows, thereby delaying discovery of the fire by passersby, accelerating the development of flashover conditions, and increasing the likelihood of the development of backdraft conditions.

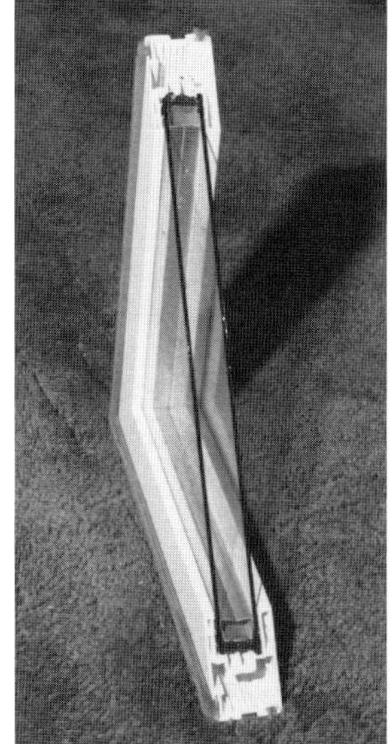

Figure 3.31 Cutaway of a double-glazed energy-efficient window showing the air-space between the panes.

There are other specialized windows designed for very specific applications. Regardless of the design or application, windows that resist opening from the outside usually can and should be broken out when a horizontal ventilation opening is needed. However, many newer windows are glazed with, or constructed entirely of, extremely resilient plastics, and they are very difficult to break. In these situations, removing the entire frame with the window intact may be quicker and more efficient for ventilation purposes (Figure 3.32).

Figure 3.32 Firefighters removing the entire window assembly.

One additional aspect of window construction must also be considered — security measures. As citizens attempt to protect themselves and their property from break-ins, a growing variety of security systems and devices are being found on the outsides and insides of windows. These devices, which have long been commonplace on commercial buildings, are now being seen on the windows of residences (Figure 3.33). Because a tool can be inserted between the bars of many security devices and still break out the window, these devices are more of a problem for forcible entry than for ventilation. However, windows covered with heavy-gauge screen may not even allow a tool to pass through, and the screen will have to be removed if that window must be opened for horizontal ventilation.

Figure 3.33 Typical window security bars.

Doors

While locked doors are often seen primarily as forcible entry problems, they also represent problems and opportunities from a horizontal ventilation standpoint. Doors present the same sorts of problems for ventilation as for forcible entry, but they also provide opportunities to make horizontal ventilation very effective. Doorway openings are usually much larger than window openings. So, for the same investment of time and effort, opening a door usually results in a bigger, and therefore more effective, ventilation opening.

Regardless of its design or location, a door opened for ventilation purposes should be blocked open or removed from the frame (Figure 3.34). Either of these measures will prevent the door from being inadvertently or accidentally closed during the ventilation operation, which could have serious consequences. For the specifics of forcing locked doors, see the IFSTA manuals **Forcible Entry** or **Essentials of Fire Fighting**.

In most cities and towns, the variety of windows in the community is rivaled only by the variety of doors. The most common types of exterior doors are the swinging door, the sliding door, the revolving door, and several different types of roll-up doors. Also of importance to horizontal ventilation are interior fire doors. Listed in the following sections are the construction features of the most common types of doors. For a more complete discussion of any aspect of building construction, refer to the IFSTA manual **Building Construction Related to the Fire Service**.

Figure 3.35 A typical single-acting, single swinging door.

Figure 3.34 A bulkhead door blocked open.

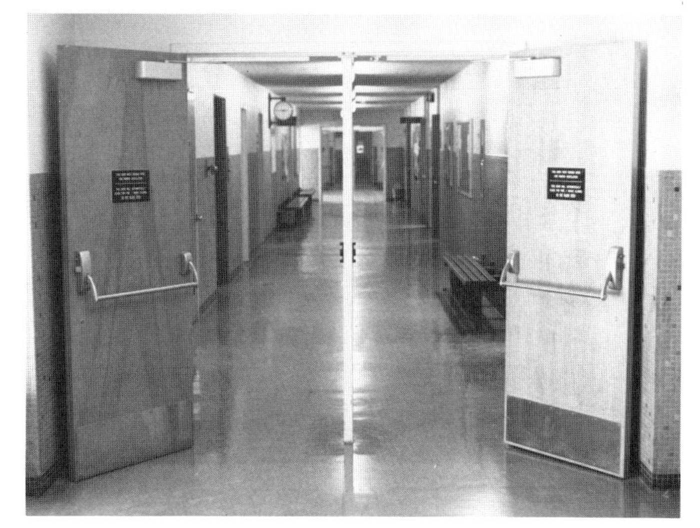

Figure 3.36 A typical set of single-acting, double swinging doors.

SWINGING DOORS

Without question, the most common type of door is the swinging door. A single door consists of one movable leaf (swinging panel) set in a rigid frame (Figure 3.35). A double door consists of two door leaves set in a single frame, and both leaves are movable (Figure 3.36). Both types of doors may be either single- or double-acting.

Single-acting doors (whether single or double leaf) swing at least 90 degrees from the frame in one direction only (Figure 3.37). They are typically hinged on one side. Double-acting doors swing 180

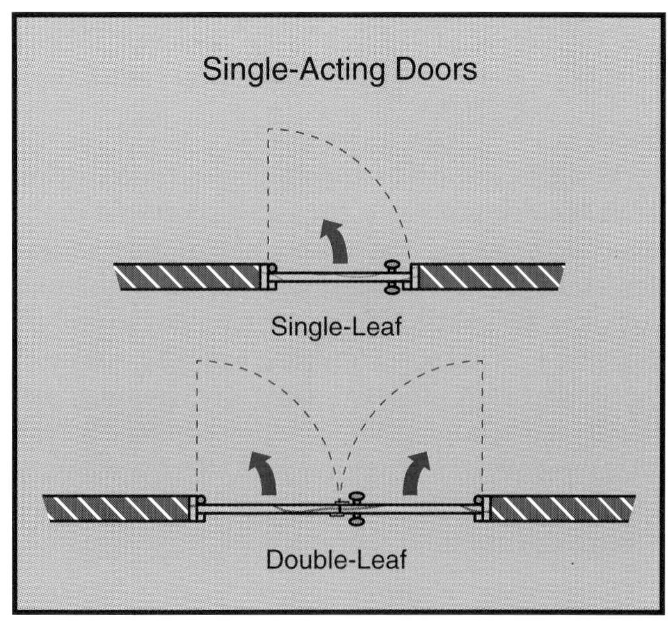

Figure 3.37 The operation of single- and double-leaf single-acting doors.

degrees from the frame in both directions (Figure 3.38). Although some have double-acting hinges, most exterior double-acting doors pivot on pins at the top and bottom rather than on hinges.

Exterior swinging doors will usually be solid-core doors and may be metal clad for security reasons. They may also be metal doors, frameless tempered-glass doors, or glass in wooden or metal frames (Figure 3.39). Interior swinging doors are often hollow-core doors, which are far less substantial than solid-core doors and are more easily forced.

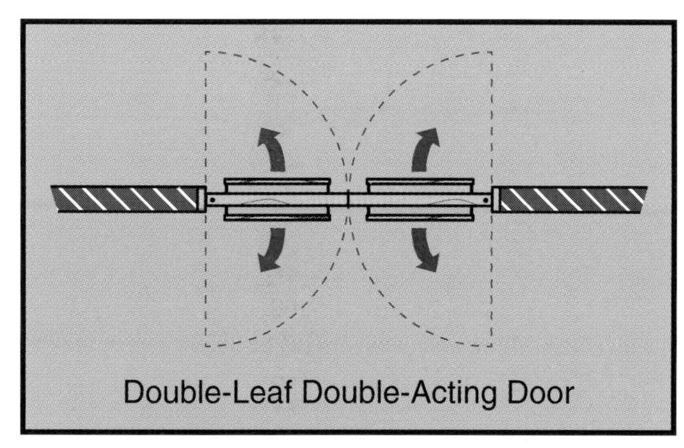

Figure 3.38 The operation of a double-acting door.

Figure 3.39 Typical exterior doors in mercantile occupancies.

SLIDING DOORS

From the standpoint of horizontal ventilation, there are three basic types of sliding doors: exterior sliding doors, interior fire doors, and interior pocket doors.

Exterior sliding doors. Exterior sliding doors are most common on storage, commercial, or industrial buildings. They are generally suspended from a horizontal track on the wall of the building (Figure 3.40). Most doors of this type are manually pushed open or closed. Because the track is level, these doors will usually remain in the position in which they were left, so blocking them open is not as critical as with other types of doors. These doors are usually metal or metal clad. They have a variety of locking mechanisms but often just a hasp and padlock. Because these doors usually cover relatively large openings, they can be very advantageous for ventilation. The other common type of exterior sliding door is the framed-glass sliding door most often found in residential units (Figure 3.41).

Figure 3.40 A commercial-type exterior sliding door.

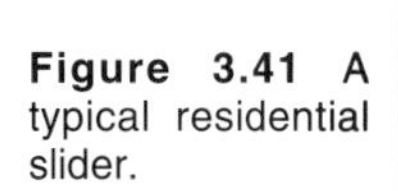

Figure 3.41 A typical residential slider.

Interior fire doors. The sliding type of interior fire door is a metal-clad, rated assembly that is suspended from a slanting track attached to the surface of a rated fire wall (Figure 3.42). Because the track is slanted, the door will remain in the closed position unless manually opened. In high-traffic areas where it would be impractical to continually open and close a fire door, the door is held in the open position by a cable connected to a fusible link at the top of the doorway (Figure 3.43). When the heat of a fire passes through the doorway, the link fuses and separates to release the cable. Gravity then rolls the door into position, closing off the opening. Before fire doors are reopened for ventilation purposes after a fire, firefighters must make sure there is no longer any danger of fire spread through that doorway. These doors may also represent a hazard to firefighters working within a building. The heat of a fire may cause the door to activate and close behind the firefighters, thereby cutting off one (and, perhaps, the only) escape route. Because they are intended to stay in place once activated, these doors tend to be quite heavy and may be very difficult to push back up the track, especially if obstructed by fallen debris.

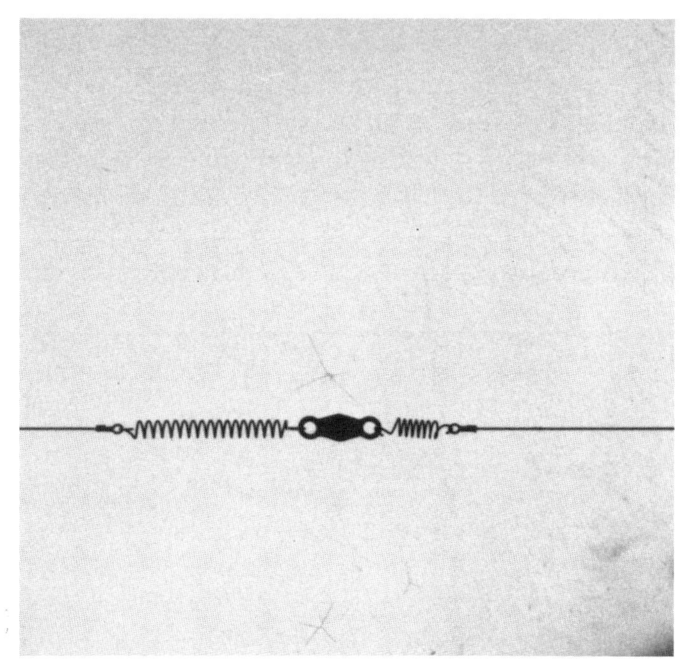

Figure 3.43 A typical fusible link.

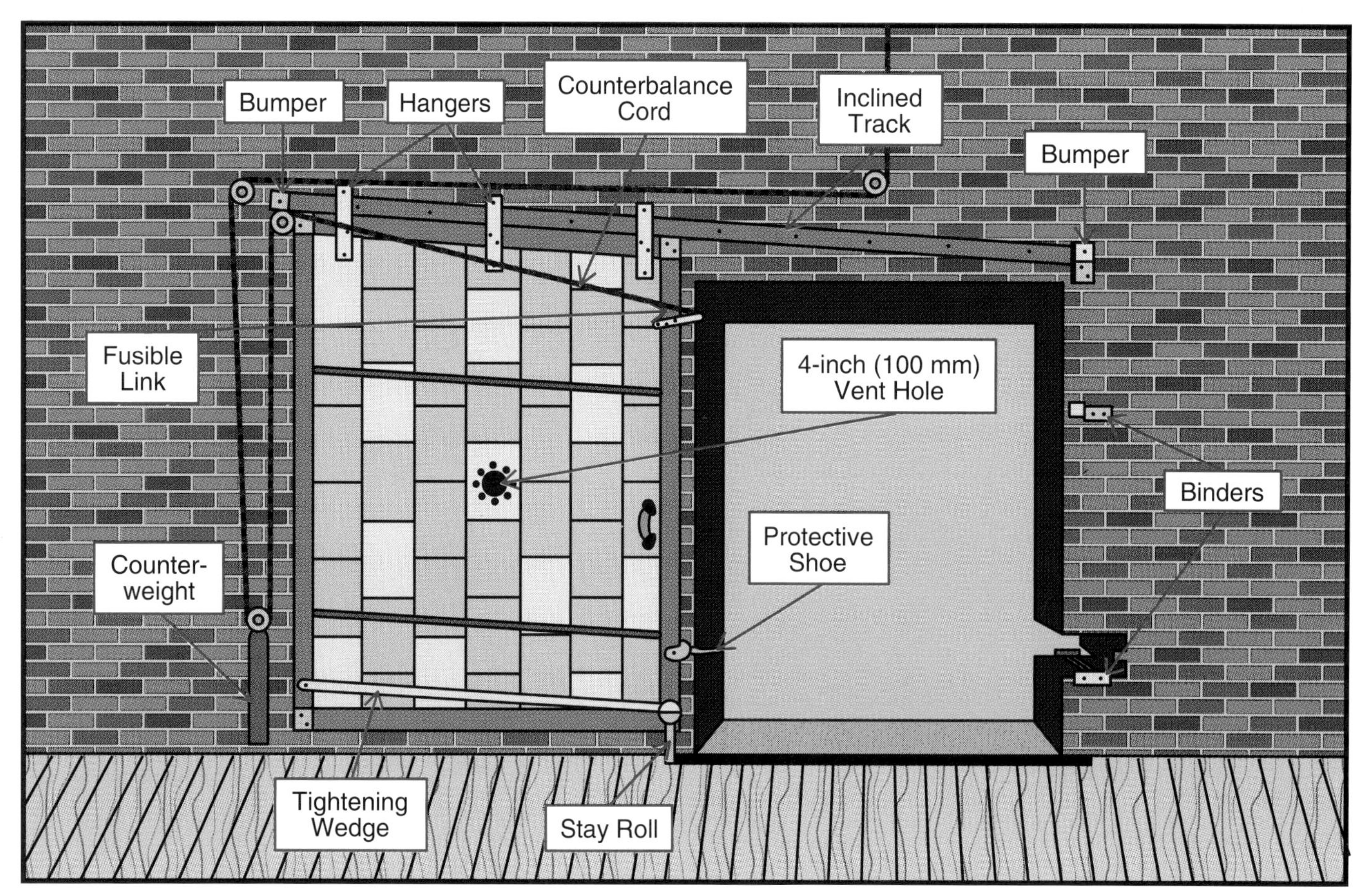

Figure 3.42 Components of a typical interior sliding fire door.

Interior pocket doors. Interior pocket doors are sliding doors that roll back into a wall, rather than rolling along one surface (Figure 3.44). They are most often lightweight, hollow-core doors used primarily as visual barriers for privacy. They are, however, sometimes used in hallways to separate various sections of an occupancy. Their importance from a ventilation standpoint is more as an impediment than as an aid to the process.

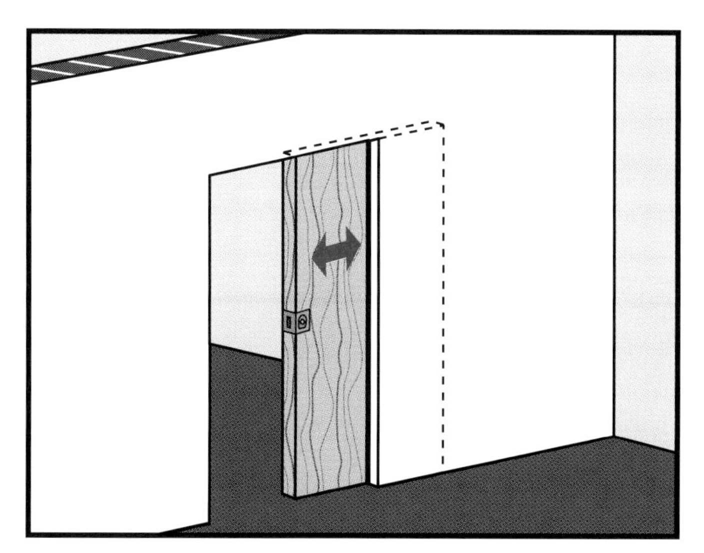

Figure 3.44 An interior pocket door.

Figure 3.45 One style of revolving door. *Courtesy of Ray C. Goad, Goad Engineering, Boonville, IN.*

REVOLVING DOORS

This type of door is usually found in large office, commercial, or apartment buildings (Figure 3.45). As with other types of doors, there is a variety of different designs, but they are more similar than different. In all cases, there is a mechanism for folding the door sections (wings) into a fixed, open position (Figure 3.46). During pre-incident planning visits, firefighters should familiarize themselves with the mechanism and the procedure for opening revolving doors. The opening can then be used effectively for horizontal ventilation.

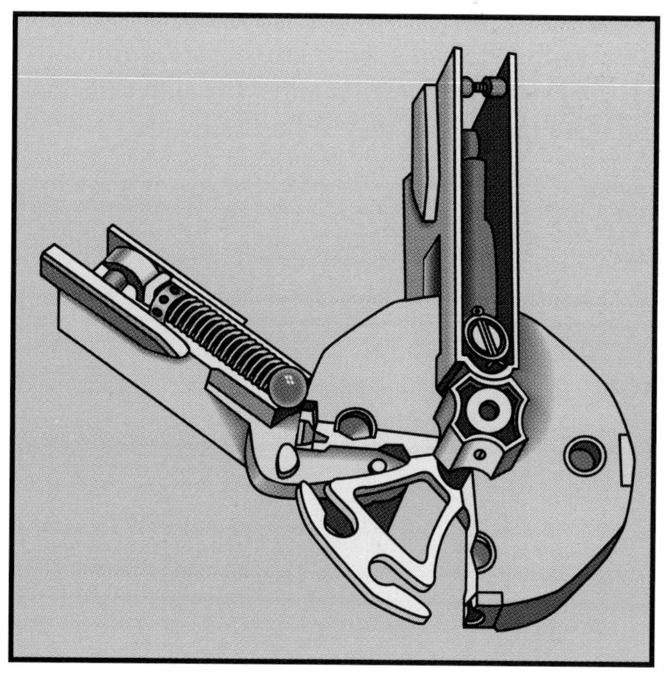

Figure 3.46 A typical mechanism for folding the door sections (wings) in a fixed, open position. *Courtesy of Ray C. Goad, Goad Engineering, Boonville, IN.*

ROLL-UP DOORS

These doors also come in a variety of types and styles. They vary from lightweight wooden or metal sectional doors, typical of residential garage doors, to heavy-duty steel rolling doors used in commercial and industrial occupancies (Figure 3.47). Some of the heavy-duty doors are also rated fire doors, held in the open position by fusible links as described previously.

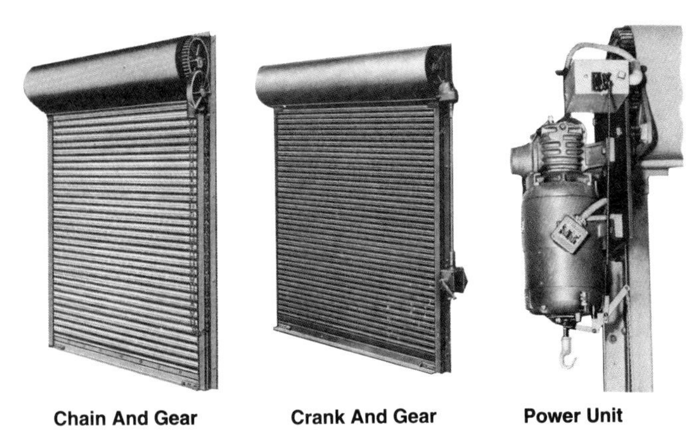

Figure 3.47 Typical commercial roll-up doors.

From a ventilation standpoint, roll-up doors are sometimes a problem because they are difficult to force open. In addition to the time involved in forcing these doors, the door components may be bent in the process. Then the door may not open fully because of the distorted metal. If the crew simply cuts a large triangular opening in the door, which is usually much faster than forcing the door, the size of the opening will be significantly smaller than the fully open door (Figure 3.48). This smaller opening will obviously reduce the volume of airflow compared to that available with the fully open door. The size of the opening can be increased to provide better access and ventilation by removing the cut ends of the slats from both sides of the opening (Figure 3.49). Roll-up doors often have a conventional swinging door adjacent to them (Figure 3.50). If so, it is usually easier to enter through this door and open the roll-up door from the inside.

Figure 3.48 A typical triangular opening cut in a roll-up door.

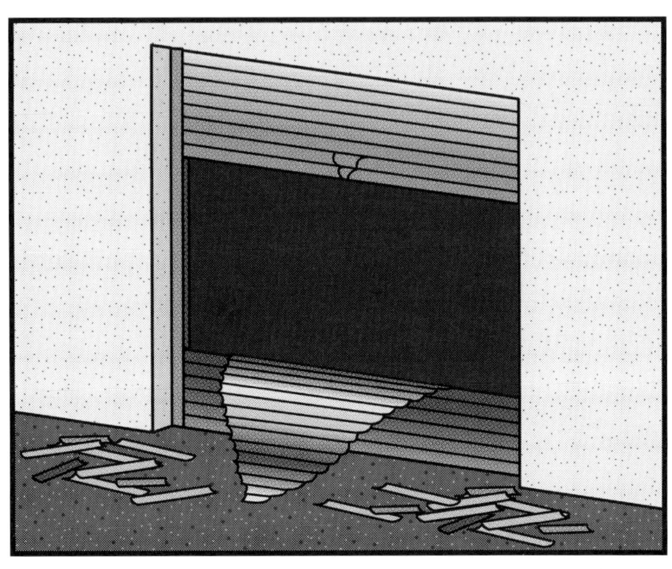

Figure 3.49 A "tepee" cut with the cut ends of the slats removed.

Figure 3.50 A conventional swinging door adjacent to a roll-up door.

ESTABLISHING AND SUPPORTING HORIZONTAL VENTILATION

Horizontal ventilation is often started when firefighters make entry into a building for search and rescue and/or fire fighting (Figure 3.51). Ideally, however, establishing horizontal ventilation will result from a more conscious decision and from a more deliberate act. While wind intensity and direction are always factors, the most important factors in deciding where to make the initial ventilation opening are that the exit opening be made as close to the seat of the fire as possible and that it be opposite the point from which attack lines will advance (Figure 3.52). If ventilation is attempted without knowledge of the location of the fire, the extent of the fire, and the arrangement of rooms and partitions in the building, uninvolved portions of the building may be jeopardized.

Figure 3.51 A firefighter initiating horizontal ventilation.

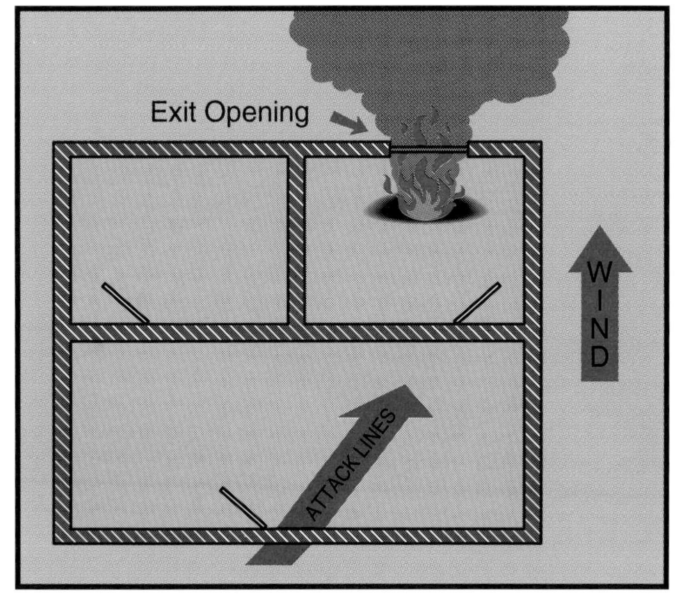

Figure 3.52 A ventilation exit opening should be as close to the seat of the fire as possible.

Just as structure fires should be attacked from the unburned side to avoid spreading the fire into uninvolved areas, horizontal ventilation should also be initiated at a point that will limit fire spread. Under ideal conditions, an exit opening should first be made on the leeward side of the building, immediately followed by an interior attack from the windward side. However, if the seat of the fire is on the windward side of the building, creating an opening on the windward side (closest to the center of the fire) would tend to spread the fire throughout the building (Figure 3.53). Under these conditions, it may be possible to counteract the effects of an adverse prevailing wind by pressurizing the building from the leeward side before creating an exit opening on the windward side (Figure 3.54).

Precautions Against Upsetting Established Horizontal Ventilation

Personnel should be careful not to block or close openings that channel fresh air into the area that is being ventilated. Established ventilation may also be upset if additional openings are made that rechannel the air currents intended for ventilating the area. Some things that can upset established horizontal ventilation are listed below:

- *Improper use of forced ventilation.* If forced ventilation is improperly applied, such as by starting the operation before the seat of the

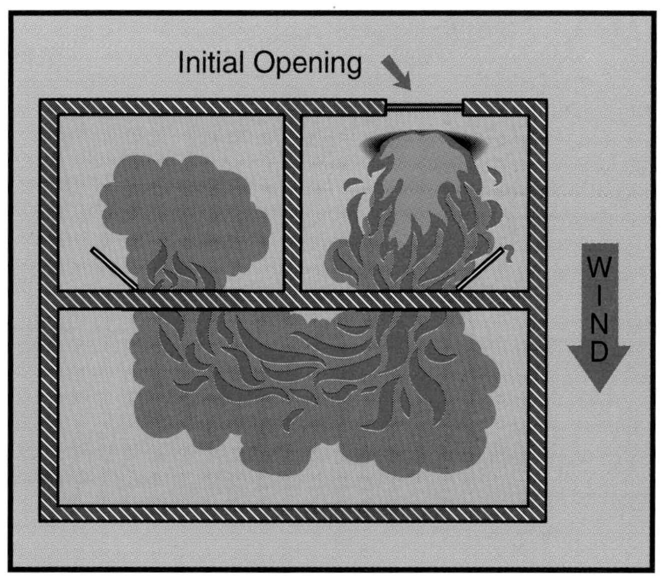

Figure 3.53 The effects of creating an opening on the windward side of a building.

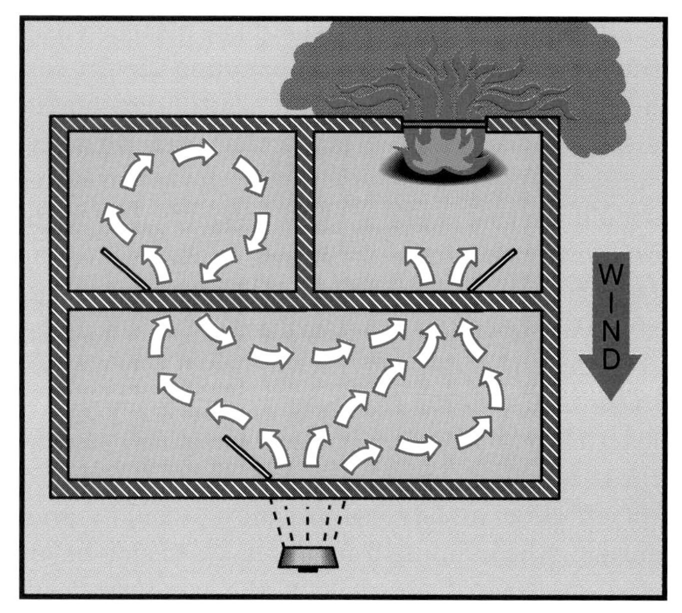

Figure 3.54 Pressurizing a building to counteract the effects of an adverse prevailing wind.

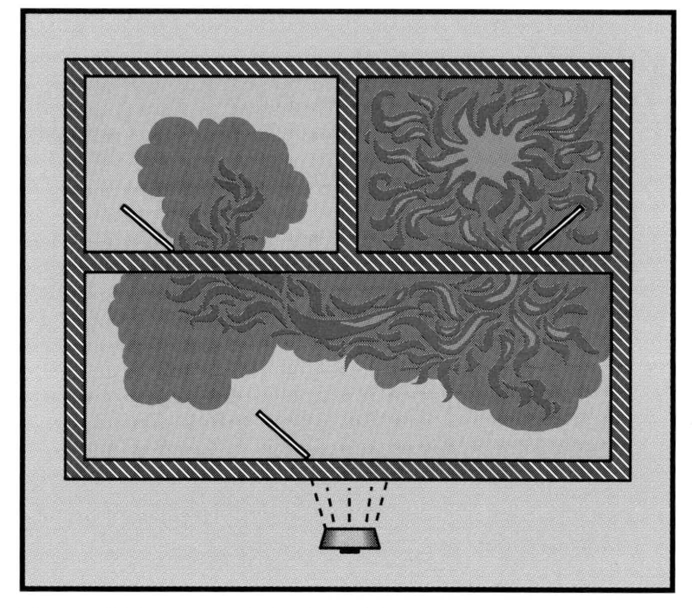

Figure 3.55 Forced ventilation intensifying and spreading a fire.

fire has been located or before attack lines are ready to be taken into the building, the fire can be intensified and can spread to uninvolved areas of the structure (Figure 3.55).

- *Inadequate control of exit openings*. Effective forced ventilation can be rendered ineffective by crews arbitrarily opening windows without being ordered to do so. With forced ventilation, a proper balance between the volume of air being introduced and the size of the exit openings must be maintained (Figure 3.56).
- *Improperly located exit opening*. With or without forced ventilation, if the exit opening is

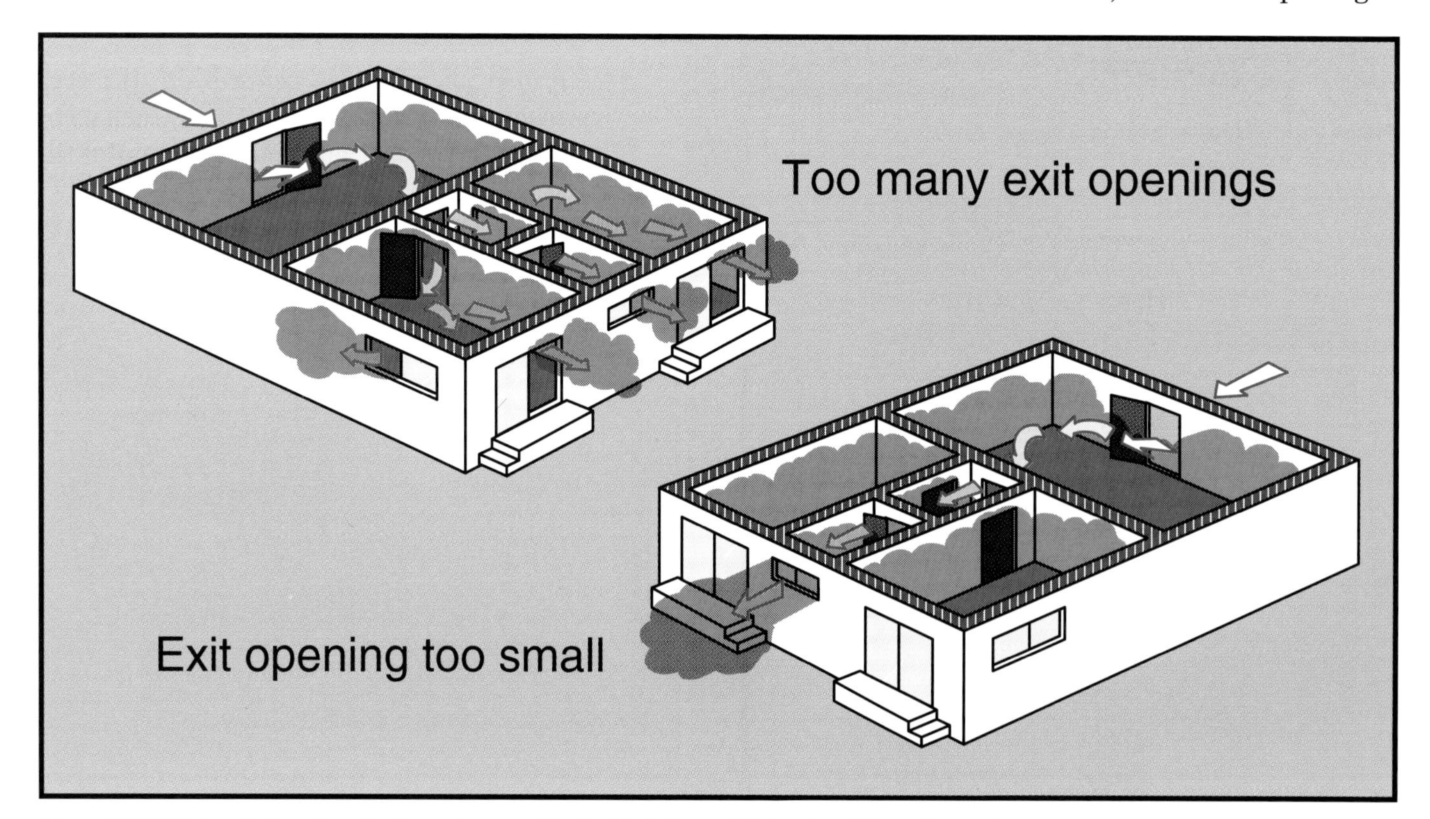

Figure 3.56 The size and number of exit openings must be kept in balance with the size of the entry opening.

improperly located so that heat and smoke are drawn into uninvolved portions of the building, escape routes may be denied to occupants, attack crews may be subjected to unnecessary punishment, and fire damage may be increased (Figure 3.57).

- *Improperly directed fire streams.* If fire streams are directed into ventilation exit openings (Figure 3.58), whether horizontal or vertical, the results can be disastrous. The natural convection currents can be reversed, causing mushrooming within the structure. Also, the additional steam generated can injure and/or drive interior crews from the structure.

- *Improper placement of salvaged contents.* If large piles of furniture or other contents are stacked in the wrong location during salvage or overhaul operations, they can have a detrimental effect on ventilation efforts. Contents stacked in hallways or too near doorways or windows can impede the flow of air into and within the structure. This condition will reduce the effectiveness of natural or forced ventilation, whether horizontal or vertical (Figure 3.59).

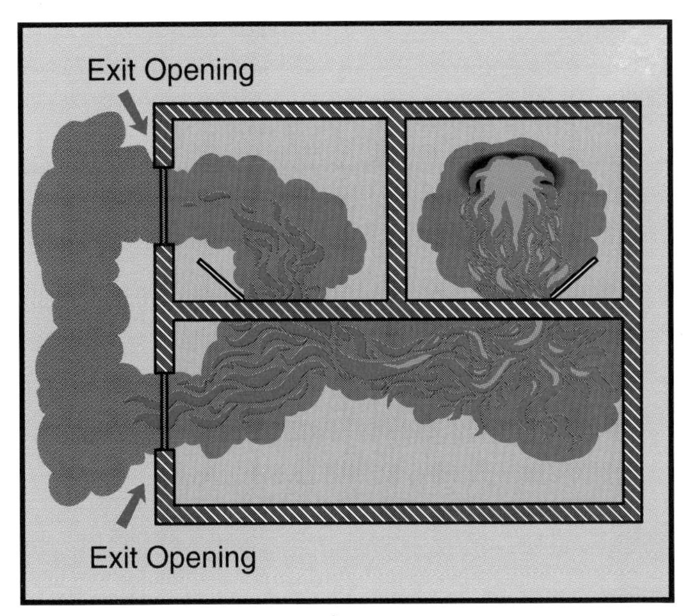

Figure 3.57 Fire being drawn to multiple exit openings.

Figure 3.59 Salvaged contents impeding the flow of air into a structure.

Obstructions To Horizontal Ventilation

Even when firefighters make good decisions about where, when, and how to ventilate a building,

Figure 3.58 Hose streams should not be directed into ventilation exit openings.

factors within the building can significantly reduce the effectiveness of the ventilation operation. Such things as the layout of the building and how the contents are arranged can create impediments to effective ventilation. Under adverse conditions, the wind can hinder the ventilation process as much as it can help it under more favorable conditions.

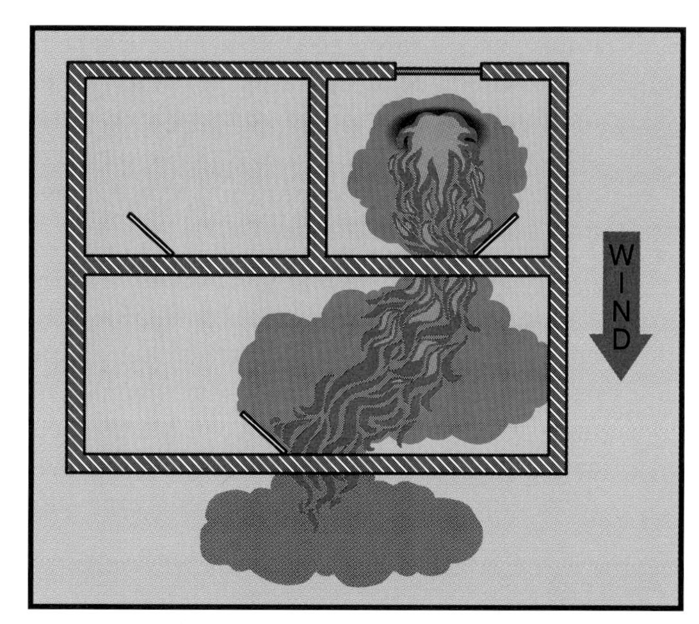

Figure 3.61 Creating an exit opening on the windward side of a building can allow the wind to spread the fire.

BUILDING CONSTRUCTION AND CONTENTS

Even though over a long period of time buildings may become completely filled with smoke that has seeped through cracks and ventilator openings, horizontal ventilation may be obstructed by walls, partitions, and stacks of stored material (Figure 3.60). If buildings contain a large number of rooms or are heavily loaded with contents, they may be very difficult to ventilate horizontally because of poor air circulation.

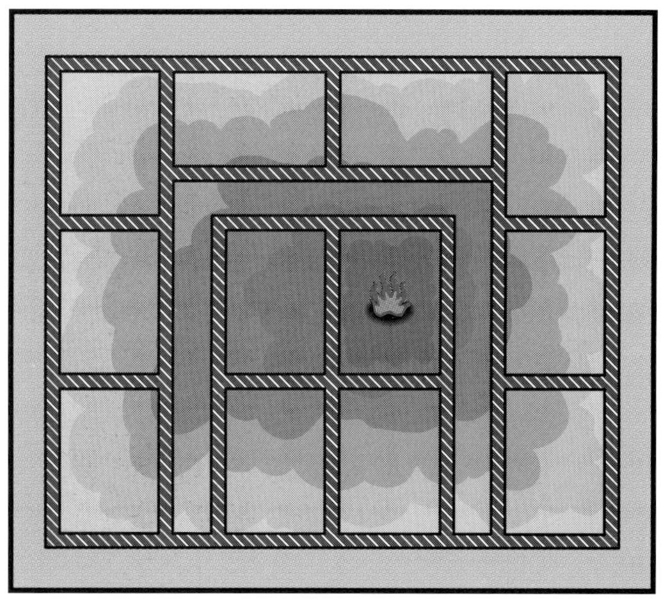

Figure 3.60 Walls, partitions, and stacks of stored material may obstruct horizontal ventilation.

Figure 3.62 Within limits, PPV can overcome the effects of an adverse wind.

WIND

Under ideal circumstances, the wind can provide all the air circulation necessary for effective horizontal ventilation. Opening the structure on the windward side and ventilating on the leeward side often works very well. However, as mentioned earlier, if the seat of the fire is on the windward side of the building, the wind would tend to spread the fire into uninvolved areas (Figure 3.61).

Depending on the size and efficiency of the blower(s) used, winds of up to 25 mph (40 km/h) may be overcome with positive-pressure ventilation (Figure 3.62). With higher wind speeds, and if vertical ventilation is impractical or impossible, delaying ventilation until after the fire is knocked down may be the only available option. If ventilation is to be delayed, firefighters must be aware of the danger of a rollover and should stay as low as possible when approaching and attacking the fire.

NATURAL HORIZONTAL VENTILATION

Of all the various methods of ventilating structures, natural horizontal ventilation is by far the most often used. Unaided by any mechanical means, this method depends entirely on the buoyancy of the smoke and the prevailing wind.

Doors

Both exterior and interior doors can play an important role in horizontal ventilation. Because exterior doors are the most likely points of entry for search and rescue or fire attack, they may become part of the ventilation operation whether intentional or not. They may have to be forced before they can be opened for ventilation or entry, and this process can sometimes be a frustrating and time-consuming task. The doors may be metal clad, have heavy security bars, or be otherwise reinforced for security purposes. Metal roll-up doors can be very difficult to force open, so cutting a large hole in them may be the most efficient choice. Before any door is forced, however, the old rule of "try before you pry" should always be followed.

Whenever any door is opened for ventilation purposes, whether for natural or forced ventilation, the opening should be maintained by blocking the door open, or by removing it (Figure 3.63). If the door should suddenly close for any reason during a critical phase of the operation, the dynamics of the situation could change dramatically, perhaps endangering occupants and/or firefighters.

Interior doors should be opened or closed in order to accomplish the objectives being pursued. These doors will have to be opened to facilitate searching the room within, but if there are open windows in the room, opening the door may have a detrimental effect on the airflow needed for ventilation. It is generally a good practice to leave interior doors closed after each room has been searched. For ventilation purposes, interior doors may be opened and closed one by one in a systematic process of ventilating an entire floor (Figure 3.64).

Figure 3.63 Bulkhead door removed from its hinges to prevent the accidental closing of the door.

Figure 3.64 Systematically opening and closing interior doors to assist in ventilating the structure.

Windows

OPENING WINDOWS

If the seat of the fire is on the leeward side of the building, the correct procedure for ventilating horizontally (cross ventilation) is to first open the top windows on the leeward side, allowing the superheated gases to escape. The next step is to then open the lower windows on the windward side to introduce replacement air into the building (Figure 3.65). When any window is opened for the purpose of ventilation, screens, curtains, drapes, or blinds should be removed because they will hinder air circulation (Figure 3.66).

If the seat of the fire is on the windward side of the building and a blower is available, the interior of the building should be pressurized before creating an exit opening on the windward side. If a blower is not available, opening windows on the windward side should be delayed until after initial knockdown of the fire.

BREAKING WINDOWS

Before breaking any window for ventilation or other purposes, do not assume it is locked. Always

Figure 3.66 A firefighter removing window coverings.

Figure 3.65 Natural cross ventilation under ideal conditions.

try to open it first. If the situation makes it necessary to break the window, break out the entire pane and clear the sash of broken shards by scraping the sash with the breaking tool. The firefighter breaking the glass should be in full protective clothing, including hand and eye protection, and should be upwind of the window. If using an axe to break the window, the firefighter should use the flat side of the blade (Figure 3.67). Regardless of which tool is used, the handle should be held higher than the blade to prevent glass from sliding down the handle. When working above ground, windows should be broken inward whenever possible to avoid the "flying guillotine" hazard to those working below.

Figure 3.67 Firefighter demonstrating the proper technique for breaking glass with an axe.

THERMOPLASTIC WINDOWS

A current trend in the construction industry is using Plexiglas® acrylic plastic and other thermoplastics in place of glass windows. Lexan® plastic, an extremely resilient glass substitute, presents a challenge for firefighters attempting ventilation during fires. Experience indicates that a circular saw with a carbide-tipped blade is most effective when entry must be made through Lexan and similar plastics (Figure 3.68). Caution must be taken in blade selection because a blade with too fine a tooth will cause the Lexan plastic to melt and the blade to bind. Conversely, a blade that is too coarse will cause the blade to slide dangerously over the cutting surface. A medium-toothed blade has been found to give the best results.

If a power saw is not available, entry might be made in one of three other ways. One is to strike the pane in the center with a sledgehammer (Figure 3.69). This procedure will not break the pane, but it may bow it enough to allow it to slip out of its frame. A second method is to use the point of a pick-head axe to score an "X" on the pane and to then strike the pane at the intersection of the "X" with the point of the pick head (Figure 3.70). The pane will usually break along the "X," and the pieces can then be

Figure 3.68 A firefighter about to use a rescue saw to cut a plastic window.

Figure 3.69 A firefighter about to strike a plastic window with a sledgehammer.

Figure 3.70 Scoring an "X" on a plastic window will make it easier to break with the point of an axe.

pulled or bent out. Either technique may not work if the pane has been heated to the point where it becomes soft and pliable. A third technique is to "freeze" the pane with CO_2 to make it brittle and to then strike it in the center with the point of a pick-head axe.

FORCED HORIZONTAL VENTILATION

Whether in horizontal or vertical ventilation operations, forced (mechanical) ventilation is primarily a means of supplementing or augmenting natural ventilation. Horizontal ventilation has traditionally been the area where smoke ejectors and/or nozzles have been used. Both of these methods are applied at the point of exit for the smoke and are almost always applied from inside the structure. More recently, blowers have begun to be used to force replacement air into a structure from the outside at the point of entry. For a more thorough discussion of forced (mechanical) ventilation, see Chapter 5, Forced Ventilation.

Using Smoke Ejectors

Smoke ejectors are capable of being set up to blow air into a structure, but that is not the purpose for which they were designed. These units were designed to withstand the heat and contamination of drawing smoke through the fan in the process of exhausting it to the outside.

While smoke and heated air may pass harmlessly through the unit, actual flame passing through it can cause serious damage. Unless they are equipped with explosion-proof motors, smoke ejectors are capable of igniting certain combustible gases encountered in fires. For these reasons, smoke ejectors are most often employed after initial knockdown of the fire and not as part of the fire attack.

Smoke ejectors are usually set up in openings on the lee side of the building. They may be suspended in window or doorway openings, set up on the floor, or elevated on some object (Figure 3.71 on opposite page). Using bolt-on telescoping legs, smoke ejectors can be made freestanding, and their height can be adjusted to the most effective level.

When smoke ejectors are placed in doorways, windows, or other exterior openings, the open area around the units must be sealed with salvage covers or some similar means to prevent churning (Figure 3.72). *Churning* is the phenomenon of smoke being blown out at the top of the opening, only to be drawn back into the structure at the bottom of the opening by the slight negative pressure (vacuum)

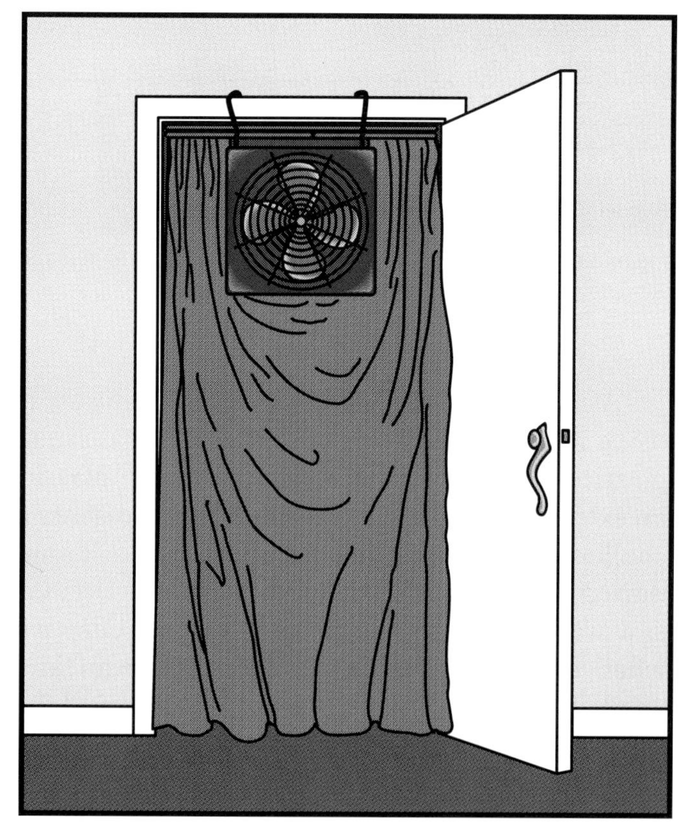

Figure 3.72 The open area around a smoke ejector sealed with a salvage cover.

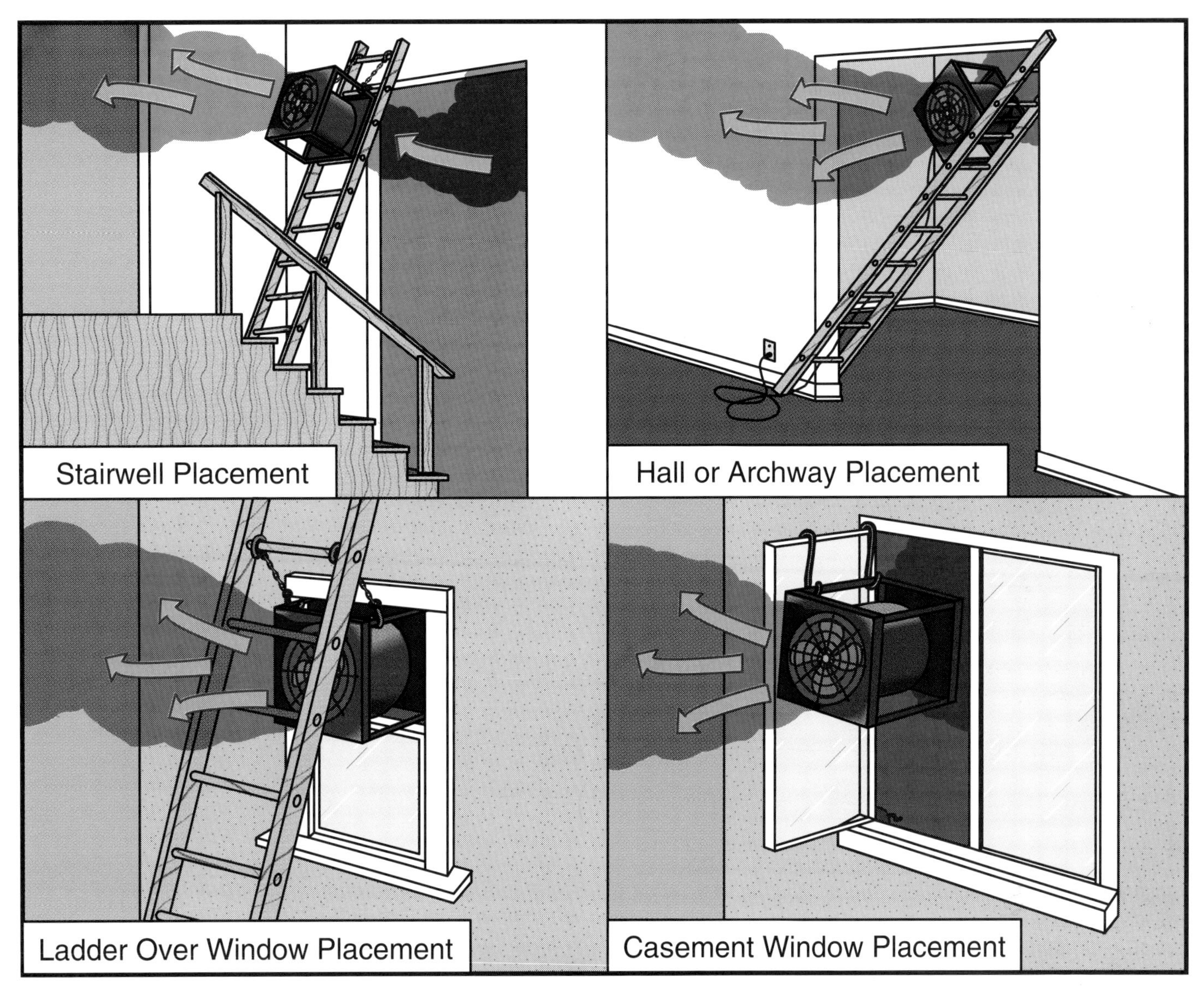

Figure 3.71 Depending on the situation, smoke ejectors may have to be positioned in a variety of openings.

created by the action of the ejector. This condition is obviously counterproductive to clearing the building of smoke.

Using Blowers

Blowers can be extremely effective adjuncts to natural horizontal ventilation. These units, usually slightly larger than smoke ejectors, are most often powered by gasoline-driven engines (Figure 3.73). Because these engines require fresh air to operate, they are obviously not designed to be set up in a contaminated atmosphere. Blowers are almost always set up a few feet (meters) outside of the point of entry into the building so they are not subject to being damaged by the fire (Figure 3.74). Properly applied, they can effectively support the efforts of the search and rescue and fire attack crews by enhancing visibility and reducing interior temperatures. For these reasons they are often used as part of the initial attack.

Blowers function by creating a slight positive pressure within the structure, thus forcing the smoke out the exit opening. A single blower is often sufficient to pressurize a building, but two blowers working in tandem can be used if needed (Figure 3.75). The keys to effective use of these units are to apply positive-pressure ventilation (PPV) as soon as the first attack crew is ready to enter and to maintain the size of the exit opening in proportion to the size of the entry opening.

While PPV can sometimes be used to help determine the location of occupants and the location and extent of the fire, some departments take

Figure 3.73 Starting a typical gasoline-powered blower. *Courtesy of Tempest Technology Corporation.*

Figure 3.74 A typical position for a blower. *Courtesy of Tempest Technology Corporation.*

Figure 3.75 Blowers set up in tandem to increase the volume of airflow into the building.

a more conservative approach and insist that these facts be known before PPV is initiated. Fire departments are encouraged to experiment with PPV in a safe, controlled training environment before developing standard operating procedures (SOPs) for the use of PPV (Figure 3.76). Once the capabilities and limitations of PPV are understood, all fire fighting personnel should be thoroughly trained in its use before PPV is implemented on the fireground.

Figure 3.76 A training officer explains what to expect when PPV is used in a training fire.

Hydraulic Ventilation

The terms *fog stream* and *spray stream* will be used interchangeably in this section. A fog or spray stream directed through a window or door opening will draw large quantities of heat and smoke in the direction in which the stream is pointed. Compared with mechanical smoke ejectors, fog streams are capable of removing two to four times more smoke, depending on the type and size of the nozzle, the angle of the spray pattern, and the location of the nozzle in relation to the ventilation opening.

A fog or spray stream directed through a window or doorway with a 60-degree pattern covering 85 to 90 percent of the opening provides the best results for ventilation. The ideal nozzle position will vary, but in most cases it should be about 2 feet (0.6 m) inside the room being cleared. Regardless of the size of the opening, the spray pattern should not exceed 60 degrees because it will lose efficiency at angles greater than that.

There are also several disadvantages to using hydraulic ventilation:

- There may be an increase in the amount of water damage within the structure.
- There will be a drain on the available water supply.
- In subfreezing temperatures, additional ice may form in the area surrounding the building.
- The nozzle operator must remain in the hostile environment during the operation (Figure 3.77).
- The operation may have to be interrupted each time the operator runs out of breathing air.
- This technique is limited to negative-pressure ventilation.

Figure 3.77 Firefighter must remain in the hostile environment during hydraulic ventilation.

Chapter 3 Review

Directions

The following activities are designed to help you comprehend and apply the information in Chapter 3 of **Fire Service Ventilation**, Seventh Edition. To receive the maximum learning experience from these activities, it is recommended that you use the following procedure:

1. Read the chapter, underlining or highlighting important terms, topics, and subject matter. Study the photographs and illustrations, and read the captions under each.
2. Review the list of vocabulary words to ensure that you know the chapter-related meaning of each. If you are unsure of the meaning of a vocabulary word, look the word up in the IFSTA **Orientation and Terminology** glossary or a dictionary, and then study its context in the chapter.
3. On a separate sheet of paper, complete all assigned or selected application and review activities before checking your answers.
4. After you have finished, check your answers against those on the pages referenced in parentheses.
5. Correct any incorrect answers, and review material that was answered incorrectly.

Vocabulary

Be sure that you know the chapter-related meanings of the following words:

- accelerating *(59)*
- adjacent *(64)*
- adjuncts *(73)*
- adverse *(65)*
- aesthetic *(55)*
- breach *(54)*
- churning *(72)*
- compromised *(53)*
- conservative *(74)*
- detrimental *(67)*
- embedded *(56)*
- enhancing *(73)*
- formidable *(54)*
- glazed *(57)*
- impede *(67)*
- impediment *(63)*
- inadvertently *(48)*
- in tandem *(73)*
- jeopardized *(65)*
- leeward *(65)*
- perimeter *(55)*
- pivot *(58)*
- pneumatic *(54)*
- prevailing *(65)*
- resilient *(59)*
- salvage *(67)*
- versatility *(50)*
- windward *(65)*

Application Of Knowledge

Describe natural or forced horizontal ventilation procedures you would use in the following situation. Refer to the listed information and Figure A below. Number and describe the most effective sequence for opening and closing doors and windows, and indicate blower or ejector placement. *(Local protocol)*

Multistory Structure
Windows: All fixed Lexan®
Exterior Doors: Metal-clad, solid-core, single-acting, in-swinging
Interior Doors: Wood panel, hollow-core, single-acting, in-swinging
Exterior Walls: Masonry veneer over wood frame
Interior Walls: Gypsum board over wood studs
Room Size: All 12 x 16 feet
Hallway: 6 feet wide

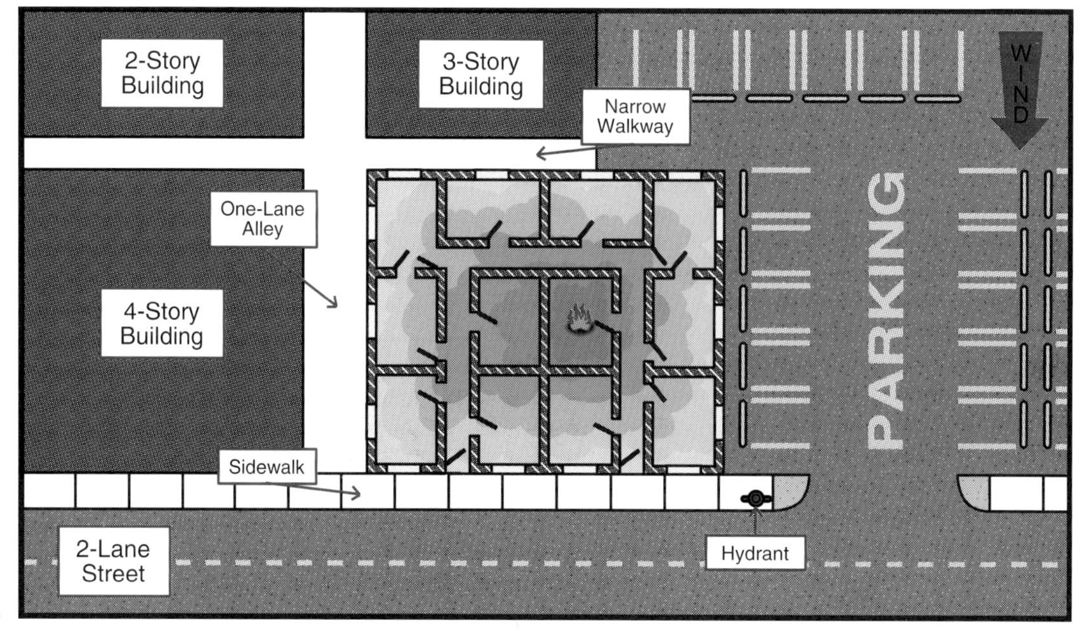

Figure A

Review Activities

1. Distinguish among the terms blower, ejector, and fan. *(49)*
2. Explain how a flexible duct is used in ventilation. *(51)*
3. List instances in which ejector/flexible duct combination ventilation would be appropriate. *(51)*
4. Identify each of the following ventilation terms:
   - fog stream *(74)*
   - forced ventilation *(65)*
   - hasp *(61)*
   - horizontal ventilation *(49)*
   - HVAC *(49)*
   - hydraulic ventilation *(74)*
   - mushrooming *(67)*
   - natural ventilation *(72)*
   - negative-pressure ventilation *(51)*
   - positive-pressure ventilation *(68)*
   - shear strength *(54)*
   - tepee cut *(56)*
   - tempered glass *(61)*
   - thermoplastic *(71)*
5. Identify each of the following construction terms:
   - below grade *(52)*
   - chipboard *(54)*
   - drywall *(52)*
   - fire resistive *(53)*
   - fire-stop *(53)*
   - gypsum board *(52)*
   - leaf (in relation to swinging doors) *(60)*
   - mullion *(57)*
   - masonry *(53)*
   - poke-through *(53)*
   - reinforced concrete *(54)*
   - sash *(57)*
   - stem wall *(52)*
   - stucco *(54)*
   - stud *(52)*
   - tilt-up concrete *(54)*
   - veneer *(54)*
   - wing (in relation to revolving doors) *(63)*
6. Describe construction features of fire walls, and explain the purposes of fire walls. *(53)*
7. List the three types of exterior walls. *(54)*
8. Discuss the pros and cons of breaching each of the exterior walls listed in Activity 7. *(55, 56)*
9. Describe construction features of each of the following types of windows:
   - fixed *(57)*
   - single- and double-hung *(57)*
   - casement *(57)*
   - horizontal sliding *(57)*
   - awning *(58)*
   - jalousie *(58)*
   - projected (factory) *(58)*
   - hopper *(58)*
10. Discuss the pros and cons of using each of the windows in Activity 9 for ventilation. *(56-59)*
11. List disadvantages of energy-efficient and thermoplastic windows. *(58, 59)*
12. Describe construction features of each of the following types of doors:
    - swinging *(60)*
    - sliding exterior *(61)*
    - sliding interior fire *(62)*
    - sliding interior pocket *(63)*
    - revolving *(63)*
    - roll-up *(63)*
13. Discuss the pros and cons of using each of the doors in Activity 12 for ventilation. *(60-64)*
14. Explain how building contents and construction can hinder the establishment of horizontal ventilation. *(64, 65)*
15. Discuss the role of the wind in establishing horizontal ventilation. *(65 & 68)*
16. Describe five external actions or conditions that can upset established horizontal ventilation. *(65-68)*
17. Compare and contrast techniques used in natural horizontal ventilation versus those used in forced horizontal ventilation. *(68 & 72)*
18. Explain the correct procedure for opening windows when using natural horizontal ventilation to ventilate a fire seated on the leeward side of the building. *(70)*
19. Explain the correct procedure for opening windows when using natural horizontal ventilation to ventilate a fire seated on the windward side of the building. *(70)*
20. Outline the correct procedure for breaking window glass for ventilation. *(70, 71)*
21. Describe effective methods of breaching thermoplastic paned windows for ventilation. *(71)*
22. Explain the effective placement of smoke ejectors for forced horizontal ventilation. *(72)*
23. Explain the effective placement of blowers for forced horizontal ventilation. *(73)*
24. Discuss the advantages and disadvantages of hydraulic ventilation. *(74, 75)*

Questions And Notes

Vertical Ventilation

This chapter provides information that addresses the following performance objectives of NFPA 1001, *Standard for Fire Fighter Professional Qualifications* (1992):

Chapter 3 — Fire Fighter I

3-9.3 Describe the advantages and disadvantages of the following types of ventilation:
(a) Vertical

3-9.6 Identify the types of tools used during ventilation.

3-9.7 Recognize the characteristics of, and list necessary precautions when ventilating at least the following roof types:
(a) Flat
(b) Shed
(c) Pitched
(d) Arched

3-9.8 Demonstrate determining the integrity of a roof system by sounding.

3-9.9 Describe how the following factors are used to determine the integrity of a roof system:
(a) Construction
(b) Visual observation
(c) Elapsed time of fire

3-9.10 Define procedures for the types of ventilation referred to in section 3-9.3.

3-9.13 Using both hand and power tools, demonstrate the ventilation of both pitched and flat roofs.

Chapter 4 — Fire Fighter II

4-9.3 Identify considerations that must be made when determining the location and size of a ventilation opening including:
(a) Availability of natural openings
(b) Location of the fire
(c) Direction in which the fire will be drawn
(d) Type of building construction
(e) Wind direction
(f) Progress of the fire
(g) Condition of the building
(h) Obstructions
(i) Relative efficiency of large vs. small openings

Safety Points

In its discussion of vertical ventilation, this chapter addresses the following safety points:

- Whenever firefighters are sent to the roof of a burning building, the vent group leader must ensure that a second means of egress is always provided.
- The vent group leader should make a conscious decision about the necessity for taking a charged hoseline to the roof.
- Firefighters should know how different types of roofs and roof coverings react when exposed to fire from below.
- Firefighters should know the strongest and weakest points of any particular type of roof in their district.
- Firefighters should be aware that hazards on the roof of a burning building, such as drop-offs behind parapet walls, open light wells, etc., can be obscured by smoke and/or darkness.
- Before deciding to step onto the roof of a burning building, firefighters should always read the roof.
- Before actually stepping onto the roof of a burning building, firefighters should sound the roof with a hand tool.
- Once on a roof, firefighters should always walk directly over the roof supports and never walk diagonally across the roof.
- As long as firefighters remain on the roof of a burning building, they should continue to read the roof for signs of collapse.
- Firefighters on the roof of a burning building should constantly monitor any wind and, whenever possible, work with the wind at their backs.
- Firefighters should cut an exit opening directly above the fire only if it does not place the firefighters in danger.
- Except for the guide man, other firefighters should be at least 10 feet (3 m) from any firefighter using a power saw.
- Firefighters should be aware that much smoke, and perhaps fire, will issue from the exit opening.
- Even though wearing SCBA, firefighters should avoid standing in smoke whenever possible.
- The vent group leader must monitor the physical condition of vent group personnel, especially in extremely hot or extremely cold weather.
- To reduce their exposure to smoke, vent group personnel should not remain on the roof after the ventilation operation is completed.

Chapter 4
Vertical Ventilation

INTRODUCTION

While many fires present characteristics common to both vertical and horizontal ventilation, only conditions specific to vertical ventilation are discussed in this chapter. The chapter also deals with establishing and supporting vertical ventilation, safety considerations, the most commonly used tools, existing roof openings, roof construction, lightweight roofs, roof coverings, and opening the roof.

Even though ventilation is not a method of fire extinguishment, firefighters can use it to fight the fire safely and more effectively and to reduce property damage. After the officer in charge of a fire (or the vent group leader) has sized up the situation, decided that ventilation is needed, determined the point at which the building should be opened, and made sure that attack and protection lines are ready, the next step is to open the building for ventilation.

One common method by which smoke and heat may be removed from a burning building is by opening the structure at its highest point (Figure 4.1). Although a smoke-filled building does not necessarily mean that there is danger of a backdraft or a flashover, releasing the contaminated atmosphere through the roof will reduce those potentials and will simplify fighting the fire.

ESTABLISHING AND SUPPORTING VERTICAL VENTILATION

Even after the fire officer has considered the type of building involved, determined the location and extent of the fire, moved personnel and tools to the roof, observed safety precautions, and selected the place to ventilate, the operation has only begun.

Figure 4.1 Firefighters opening a structure at its highest point.

Vertical ventilation involves all of these factors and many other precautions and procedures that the officer in charge must consider if the operation is to be successful. Prior to and during the actual opening of the roof, the officer in charge should give consideration to the following items:

- Maintaining the safety of the firefighters
- Providing a second means of egress from the roof
- Having charged attack and protection lines ready
- Observing wind direction and intensity

- Noting any obstructions or dead loads on the roof
- Reading (observing) the roof continually
- Locating the seat of the fire
- Coordinating with attack companies
- Using existing roof openings whenever possible
- Cutting one large opening, not several small ones
- Enlarging the original opening instead of cutting an additional hole if more ventilation is needed
- Evaluating roof construction type and condition
- Keeping track of elapsed time into the incident

Vertical ventilation is intended to allow the fire and its gases to escape harmlessly into the atmosphere, especially in potential backdraft situations. In performing vertical ventilation, firefighters should seek to give the smoke and fire gases the most direct path out of the structure that is safely possible, without spreading the fire or interfering with occupants exiting the building. Ideally, the ventilation exit opening should be made directly over the seat of the fire. This area is the preferred location, but *only if it is judged to be a safe one* (Figure 4.2).

Figure 4.2 Making the exit opening directly over the seat of the fire may not be safe because of heavy fire involvement. *Courtesy of Joseph J. Marino.*

Roof features can either help or hinder ventilation operations. Some features, such as solar panels, are obstructions. Others, such as air-handling units, process vents, machinery vents, and dust-collection units, may lend themselves to aiding ventilation efforts under certain conditions. A thorough analysis of all roof features should be made while the structure is inspected during pre-incident planning (Figure 4.3). If it is determined that a particular feature cannot be used for fire-related ventilation purposes, alternative methods of ventilation can and should be devised and written into the pre-incident plan.

Figure 4.3 A firefighter checking roof features.

VERTICAL VENTILATION SAFETY

There are numerous ways in which firefighter safety can be compromised during vertical ventilation operations. Vision on rooftops is often obscured by smoke and/or darkness. The tools that firefighters use to cut ventilation holes in roofs can be dangerous for the operator and for fellow firefighters nearby. Fire-weakened roofs may collapse — sometimes without warning. Creating an opening through which the products of combustion may escape can expose firefighters to extremely toxic substances and to a considerable amount of fire. These and other conditions encountered during vertical ventilation operations make the following discussion of firefighter safety especially important.

Vertical Ventilation Hazards

The most important safety consideration for firefighters actually performing vertical ventilation is the fact that they are often standing above the fire on a structure that is being weakened by fire below them (Figure 4.4). Therefore, they must become familiar with the various types of construction and the effects that fire exposure is likely to have on each. Before stepping onto any roof, firefighters should read (observe the condition of) and sound (test the condition of) the roof, and they should continue to do so as long as they are on the roof (see more detailed descriptions in Reading The Roof and Sounding The Roof sections). This procedure minimizes the chances of firefighters getting on or remaining on a roof that is structurally unsafe. Smoke may reduce visibility and make reading the roof difficult. In these cases, firefighters should sound the roof with a hand tool as they advance. Rain, snow, and ice may also interfere with accurately reading a roof, will increase the live load on the roof, and will make the surface slippery. As soon as ventilation is completed, or when conditions indicate that the roof is unstable, firefighters should immediately leave the roof.

Figure 4.4 Firefighters on fire-weakened roof.

While there are numerous hazards involved in vertical ventilation operations, the primary ones are that personnel must work above ground (often far above) on sometimes steeply sloped surfaces and that they must face the possibility of roof collapse (Figure 4.5). Personnel must beware of falling onto roofs from high parapet walls (Figure 4.6), of falling off roofs because of tripping over low parapets, and of poor footing on steep roofs. They must also beware of the toxic products of combustion released through their ventilation efforts and of the dangers of fire spread beneath them.

Figure 4.5 In vertical ventilation operations, firefighters must consider the possibility of roof collapse. *Courtesy of Harvey Eisner.*

Figure 4.6 A firefighter inspects a high parapet wall.

Firefighters should *never* get on a roof wearing anything less than full protective clothing, SCBA, and a PASS device. As a minimum, the officer in charge of the vent group should also be equipped with a portable radio (Figure 4.7).

Figure 4.7 A properly equipped vent group.

Getting Personnel To The Roof

Because truck-company personnel (or those responsible for performing truck functions in the absence of an actual truck company) are responsible for accomplishing vertical ventilation when it is deemed necessary, they should be trained to automatically determine the best means of access to the roof of a fire building. Access to the roof may be accomplished with ground ladders or aerial devices if these are not already being used for rescue operations. A good standard procedure is to automatically position the aerial device for best access to the roof at every fire and every training drill (Figure 4.8). When established as a standard operating procedure, this practice can save valuable time in ventilation and attack operations. In those cases where firefighters cannot reach the roof by using fire department ladders, they may use the building's fire escapes or those in adjoining buildings if they can be used safely (Figure 4.9).

CAUTION: Whenever firefighters ascend to the roof, by whatever means, a secondary means of egress from the roof *must* be provided. This requirement is most often fulfilled by placing ladders on two sides of the building (Figure 4.10).

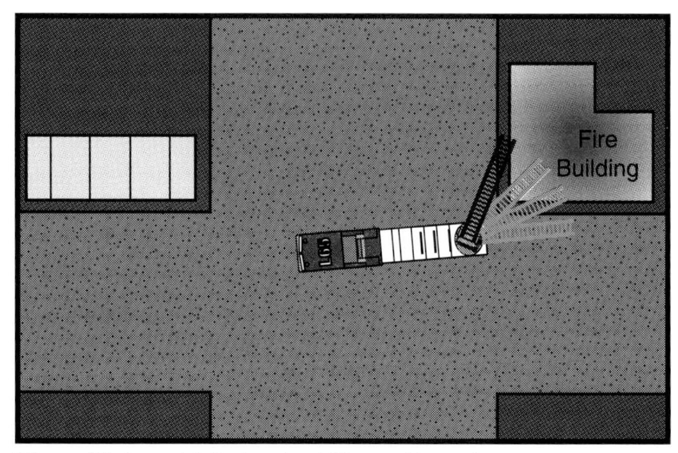

Figure 4.8 An aerial device should be positioned for best access to the roof.

Figure 4.9 A firefighter using a building's fire escape to access the roof.

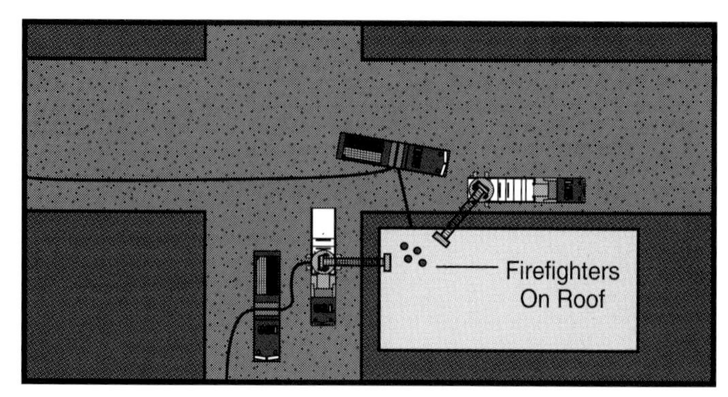

Figure 4.10 Fire department ladders can be placed on two sides of a building to give a primary and secondary means of egress from the roof.

Reading The Roof

To read a roof means to recognize certain construction features and other telltale signs that can warn the firefighter of potentially unsafe conditions. By being familiar with standard local roof and building construction design practices, firefighters can learn to "read" a roof while observing it from a position of safety such as from a ladder or an aerial device (Figure 4.11).

Construction features that can be read before venturing onto a roof include the age of the building, the type of roof structure, the location and most likely direction of supporting members, the type and condition of the roof coverings, and the existence of heavy structures and/or other dead or live loads on the roof (Figure 4.12). Because vents, skylights, and other features penetrate a roof *between* the rafters, they offer another clue to the location of roof supports. Also, the weathering of roof coverings over time will often reveal where rafters are located and the direction in which they run.

Figure 4.11 A firefighter reading a roof from an aerial ladder.

Figure 4.12 A typical example of a dead load that can be "read" before stepping onto a roof.

Other factors that firefighters should look for before getting onto a roof include:

- Sagging roof surface
- Roof vents that appear to be getting taller (indicates that the roof is sagging)
- Melted ice/snow in one area
- Melted/bubbling roofing tar
- Fire/smoke coming from roof vents
- Heavy dead or live loads
- Drop-offs due to varying roof elevations

All of these signs and conditions will aid in deciding whether the roof is stable enough to walk on or whether ventilation work *must* be done from an aerial device or other position of safety such as from the roof of an adjoining building (Figure 4.13). If the roof appears stable, the items observed while reading the roof can help identify the strongest and weakest areas of the roof and the safest routes of travel on it. If the roof appears unstable, the IC must be notified immediately so that withdrawing interior crews can be considered.

Sounding The Roof

Before stepping off a ladder, a parapet wall, or another place of safety onto the roof of a building in which a fire is burning, firefighters should *always* sound the roof by striking the roof surface with the blunt end of a pike pole, rubbish hook, or axe (Figure 4.14). When struck by a tool, a roof will feel

Figure 4.13 Firefighters working from an adjoining building.

Figure 4.14 Firefighters sounding a roof from a parapet wall.

solid over structural supports, and the tool will tend to bounce off the surface. Between the supports, the roof will tend to feel softer and less rigid. The roof will also *sound* solid when struck over a rafter or beam and will produce a more hollow sound when struck between the supports. By practicing on structurally sound roofs, firefighters can learn to recognize the difference in the feel and the sound of supported and unsupported areas of a roof (Figure 4.15).

WARNING

If structural supports cannot be located by sounding a roof, firefighters should not step onto the roof. The fire may have burned away the supports, and the roof may be in imminent danger of collapse.

Figure 4.15 Firefighters practicing on a structurally sound roof.

As mentioned earlier, as long as firefighters remain on the roof of a fire building, they should continue to read the roof for changes in its stability. In addition, they should also continue sounding the roof whenever they move about on it.

Working On The Roof

In general, the strongest points of any roof are directly over roof supports and where the roof intersects with the building's outside walls. The weakest points are between the supports (Figure 4.16). Firefighters can reduce the risk of falling through a fire-weakened roof by walking only over roof supports. Because these supports always run perpendicular to the outside walls, firefighters should *never* walk diagonally across a roof (Figure 4.17).

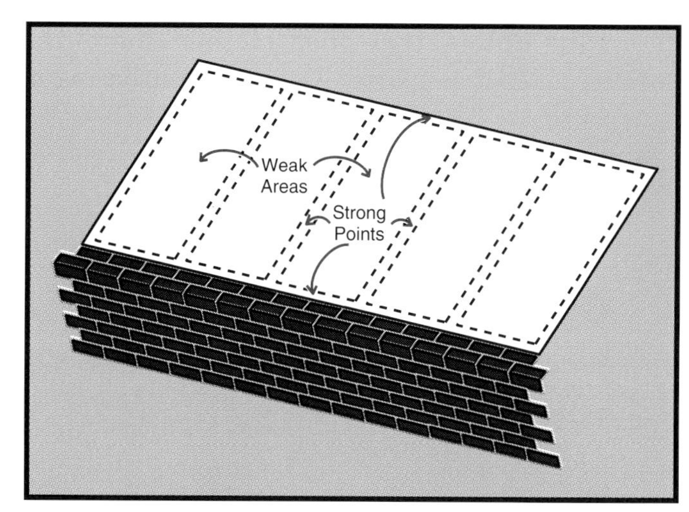

Figure 4.16 Typical weak areas and strong points of a roof.

In addition, fire-weakened roofs will sometimes fail under the weight of a heavy snow accumulation. They may even fail under the weight of a single firefighter stepping onto the surface because the weight of the firefighter is concentrated in one spot and is added suddenly. To reduce this possibility, firefighters should spread their weight over a greater area by working from a roof ladder (Figure 4.18). This technique will give the firefighter much more secure footing on pitched roofs, and because it distributes their weight, it can also be applied to flat roofs in the same way ladders are used for ice rescues.

CAUTION: Working on a roof that is known to be weakened is very risky and is *not recommended.*

Figure 4.17 Firefighters should never walk diagonally across a roof.

Figure 4.18 Working from a roof ladder spreads a firefighter's weight over a greater area.

The officer in charge of the vent group must size up the roof very carefully and give the incident commander the benefit of his or her best judgment so that the risks and benefits of ordering personnel onto the roof can be weighed. Whenever there is any doubt about the structural integrity of a roof, any vertical ventilation procedures attempted *must* be limited to those actions that can be done from an aerial device or another position of safety (Figure 4.19).

Figure 4.19 A firefighter working from an aerial ladder.

Smoke Control And The Safety Hoseline

Ideally, ventilation is performed with the wind at the firefighters' backs. However, this is not always possible such as when the seat of the fire is at the windward end of a building. Wind that may be fanning the fire may also push the fire and smoke toward the ventilation team (Figure 4.20). Under these circumstances, it may be necessary to have a charged hoseline on the roof with the ventilation team for their protection. The need for a charged hoseline on a roof should be based on the need to protect the vent team. Pertinent safety considerations are wind direction and intensity, size of the roof, type of roof covering, and the time needed to cut the hole.

A charged hoseline may be necessary to push heat and smoke away from the vent team while the team cuts an exit opening on the roof of a large-area commercial building. A hoseline may not be required on the roof of a single-family residence roof because firefighters can cut the hole and exit the roof in a relatively short time (Figure 4.21).

The hose stream may also be used to cool the thermal column rising from the ventilation exit opening. When the hoseline is used for this purpose, it is important that the spray stream be directed horizontally or at a slight upward angle across the opening but *never into the exit opening itself* (Figure 4.22). If the hose stream is directed into the opening, it will counteract the natural

Figure 4.20 Wind may push the fire and smoke toward the ventilation team.

Figure 4.21 A charged hoseline may not always be required on a roof.

Figure 4.22 A hose stream can be used to cool a thermal column, but the stream must not be directed into the ventilation hole itself.

convection currents and push steam and smoke down into the building and toward the firefighters working inside. The safety hoseline is also useful for putting out spot fires that may occur in combustible roof coverings.

CAUTION: While a charged hoseline may be a useful and necessary item to have on the roof during ventilation operations, firefighters must realize its limitations and not become overconfident and careless because it is at hand.

VERTICAL VENTILATION TOOLS

Just as in horizontal ventilation, almost any forcible entry tool may also be used for vertical ventilation, and most of them are. However, some tools are better suited to this kind of work than others. The specific applications of those tools more commonly used for vertical ventilation are discussed in the following sections.

Cutting Tools

Using any cutting tool can be a somewhat dangerous operation, both for the operator and for others nearby. Cutting tools can be especially hazardous in vertical ventilation operations because they must often be used on steeply pitched roofs, while working from roof ladders, and for cutting materials that were never intended to be cut. One of the most important safety procedures to remember is that a safe distance should always be maintained between the operator and other firefighters. While local protocol must be followed, a clear space of at least 10 feet (3 m) in all directions from anyone using a cutting tool for vertical ventilation should be maintained. One exception to this rule is the use of a guide man or backup who watches where the operator is going, clears the path of obstructions, and provides safety and support during cutting operations (Figure 4.23). The three most common cutting tools used for vertical ventilation are the pick-head axe, rotary saw, and chain saw.

PICK-HEAD AXE

When power saws are unavailable or inoperative, a pick-head axe can be used effectively to open roofs, but doing so is a much more strenuous operation (Figure 4.24). Cutting should be done as close to the rafters as possible to minimize the tendency of the roof surface to deflect when struck (because of the springiness of the surface between the rafters) and to keep the axe from bouncing. Using a sharp axe also makes the job easier, but at best, opening a roof with a pick-head axe is a very arduous procedure. The axe can also be used to mark the roof covering where cuts are to be made with a power saw (Figure 4.25). Another use of the pick-head axe is scraping away gravel from areas to be cut.

ROTARY SAW

As with any gasoline-powered tool, the rotary saw should be started and run briefly at the ground

Figure 4.24 A firefighter using a pick-head axe to open a roof.

Figure 4.23 A guide man is used to watch where the operator is going and to clear the path of obstructions.

Figure 4.25 A firefighter marking the roof covering where saw cuts are to be made.

Figure 4.26 Power saws should be started on the ground and run briefly before being sent aloft.

or street level, and then shut off, before it is sent aloft (Figure 4.26). Also at the street level, the blade should be changed if the material to be cut requires something other than the multipurpose blade normally carried on this type of saw. When using this or any other power tool, the firefighter should wear full protective clothing (especially eye and ear protection).

After marking the perimeter of the ventilation hole with an axe, the operator should place the saw flat on the roof surface (Figure 4.27), rev it up to cutting speed, and then slowly rock the saw forward until the blade contacts and cuts the roof covering to the required depth (Figure 4.28). The saw should continue cutting as it is being drawn back toward the operator, who moves backward one step at a time. When the saw blade comes into contact with a rafter, the saw should be rocked back slightly to decrease the depth of cut until the blade is clear of the rafter (Figure 4.29).

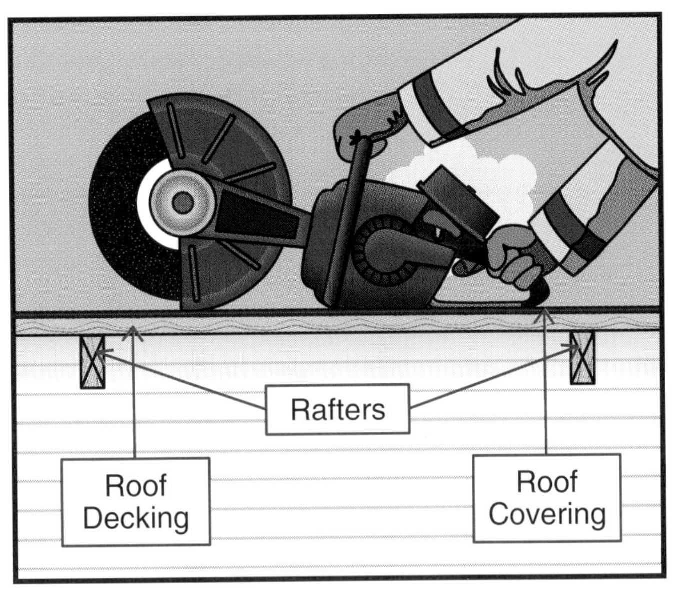

Figure 4.27 A rotary saw in position to start the first cut.

CHAIN SAW

All of the safety precautions that apply to the rotary saw also apply to the chain saw. Because of its versatility and safety, the chain saw is the preferred cutting tool among many truck-company personnel. Chain saws are usually equipped with carbide-tipped chains to allow cutting through nails and lightweight metal components. They may also be equipped with an adjustable depth gauge to reduce the chances of cutting through rafters (Figure 4.30).

Figure 4.28 The saw cutting through the roof covering and sheathing.

Figure 4.30 A chain saw equipped with an adjustable depth gauge. *Courtesy of Cutters Edge.*

Figure 4.29 The saw being rocked back slightly to avoid cutting the rafter.

Figure 4.31 A firefighter making a roof cut using only the last few inches at the tip of the cutting bar.

The procedure used with chain saws is very similar to that used with rotary saws: The chain speed is increased to cutting speed before engaging the material to be cut, and the saw is drawn back toward the operator. Unlike the technique used when cutting logs, limbs, or structural members, which uses the area of the bar closest to the saw motor, roof cuts are made using only the last few inches at the tip of the cutting bar (Figure 4.31). One of the advantages of the chain saw is said to be that it allows the operator to have a better "feel" for what is being cut, thereby reducing the risk of accidentally cutting through rafters.

Chain saws are generally safer to use than rotary saws because they do not twist in the operator's hand when revved up as rotary saws tend to do. Many chain saws also have some form of chain brake to instantly stop the movement of the chain if it becomes jammed and "bucks" out of the material being cut (Figure 4.32). As with any cutting tool, however, careful and responsible

Figure 4.32 A chain saw with a typical chain brake.

operation by well-trained personnel is still the best safety device.

BURNING BAR

These ultrahigh-temperature cutting devices are capable of cutting through virtually any metallic, nonmetallic, or composite material (Figure 4.33). They will cut through materials that cannot be cut with an oxyacetylene torch, such as concrete or brick, and they will cut through metals much faster than will an oxyacetylene unit. The torch feeds oxygen and up to 200 amperes of electrical power to an exothermic cutting bar that produces cutting temperatures in excess of 10,000°F (5 538°C). The cutting bars or rods range in size from ¼ to ⅜ inch (6 mm to 10 mm) in diameter and from 22 to 36 inches (550 mm to 900 mm) in length.

Figure 4.33 A firefighter using a burning bar. *Courtesy of Tom McCarthy, Chicago (IL) Fire Department.*

Stripping Tools

Once the roofing has been cut around the perimeter of the exit opening, the roof covering must be stripped back to expose the sheathing, unless a louver vent is being created. The sheathing, whether plywood decking or individual planks, can then be stripped from the rafters. While some departments have created their own specially designed stripping tools, the most common ones are the pick-head axe, the pike pole, and the rubbish hook. A sledgehammer performs the equivalent function on tile or slate roofs.

PICK-HEAD AXE

The versatile pick-head axe can be used to strip the roof covering from the sheathing and then the sheathing from the rafters. A firefighter is first positioned on each side of the cuts. Facing leeward and working as a team, the firefighters insert the picks of their axes into the leeward crosscut and pull the roof covering toward them as they back away (Figure 4.34). Then, they repeat the process to pull the sheathing.

PIKE POLE

Pike poles are used to strip roofing in much the same way as pick-head axes are used. The hook is

Figure 4.34 Firefighters using pick-head axes to pull the roof covering.

inserted into the leeward crosscut, and the roofing is pulled back. Pike poles have longer handles than axes, and their length allows firefighters to position themselves farther away from the point where heat, smoke, and perhaps fire may be issuing forth. Also, if there is a ceiling below the roof, its removal will be necessary in order to release the contaminants, and the handle of the pike pole is an excellent tool for this purpose (Figure 4.35).

RUBBISH HOOK

The rubbish hook is used in exactly the same way as the pike pole when stripping roofing. One further advantage of the rubbish hook is that it has two hooks instead of one, giving the tool twice the purchase on the material being pulled. Another advantage is the D-handle, which allows the operator a much more positive grip and therefore a stronger pull (Figure 4.36).

SLEDGEHAMMER

While not normally thought of as a stripping tool, the sledgehammer can be used on tile- or slate-covered roofs in much the same way that the pick-head axe is used with other roof coverings (Figure 4.37). Short, controlled blows with the sledgehammer will shatter the tiles without the head of the tool penetrating deeply enough to become lodged between the sheathing boards. If a sledgehammer is not available, a flat-head or pick-head axe can be used for this purpose.

Figure 4.36 A firefighter using a rubbish hook to pull roof covering.

Figure 4.35 A firefighter using the handle of a pike pole to push down the ceiling below.

Figure 4.37 A firefighter using a sledgehammer to shatter roof tiles.

EXISTING ROOF OPENINGS

Existing roof openings may be in the form of scuttle hatches, bulkheads, skylights, monitors, or light and/or ventilation shafts. With the exception of light and ventilation shafts, all such openings are likely to be locked or secured in some manner against entry.

Scuttle Hatches

Scuttle hatches are normally square or rectangular metal-covered hatches that provide entrance from a roof into the attic or cockloft and give access to the top floor by a ladder (Figure 4.38). If the scuttle hatch is to be opened for ventilation purposes, the ceiling directly beneath the hatch may require removal, as well as any walls enclosing the scuttle access.

Figure 4.38 A firefighter opening a typical roof scuttle hatch.

Bulkheads

Bulkheads are the structures that enclose the tops of stairways that terminate on the roof (Figure 4.39). They usually have a metal-clad exterior door of standard size. Bulkhead doors may be forced open in the same manner as other doors of similar type.

Figure 4.39 A typical rooftop bulkhead.

Skylights

Skylights covering roof openings (Figure 4.40) as well as those covering atria (Figure 4.41) may be used effectively to ventilate heat and smoke. If possible, skylights should be removed instead of being broken. The flashings should be pried loose on all sides and the skylight lifted off completely. An alternative is to pry loose three sides and use the fourth side as a hinge (Figure 4.42). Skylights equipped with thermoplastic panels or ordinary window glass can act as automatic vents because the temperature of a fire will melt the plastic or cause the glass to break and fall. However, skylights equipped with wired glass may have to be removed for ventilation purposes as they resist being broken (Figure 4.43).

Figure 4.40 A typical bubble skylight on a commercial building.

Figure 4.41 A typical atrium skylight.

Figure 4.42 A skylight pried loose on three sides and the fourth side used as a hinge.

Figure 4.43 A typical skylight equipped with wired glass panels.

Monitors

Monitor vents are square or rectangular structures that penetrate the roofs of single-story or multistory buildings to provide additional natural light and/or ventilation (Figure 4.44). A monitor may have metal, glass, wired glass, or louvered sides. During a fire, those with glass sides provide ventilation when the glass breaks. If the fire has not yet generated enough heat to break the glass, the glass will have to be broken or removed by firefighters. Monitors with solid walls should have at least two opposite sides hinged at the bottom and held closed at the top with a fusible link that will allow them to open automatically in case of fire (Figure 4.45).

Figure 4.44 A common type of monitor vent.

Figure 4.45 A monitor with a side panel open.

Light And Ventilation Shafts

Light and/or ventilation shafts usually do not require opening or enlarging for fire ventilation purposes. They may, however, represent a hazard to firefighters if the shafts are not protected by a parapet wall (Figure 4.46). If vision is obscured by smoke and/or darkness, firefighters can fall into unprotected shafts — another reason why sounding the roof is especially important.

Figure 4.46 A typical light shaft in an older building. *Courtesy of Edward Prendergast.*

ROOF CONSTRUCTION

The extent to which firefighters are able to safely and efficiently ventilate a building through its roof will depend to some degree on the firefighters' knowledge of roof construction. Construction practices and materials vary in different regions, so firefighters are encouraged to visit buildings under construction in their response districts to become familiar with local conditions.

Pitched Roofs

The most common examples of pitched roofs are the gable, lantern, hip, shed, bridge truss, mansard, modern mansard, gambrel, sawtooth, and butterfly styles. The degree of slope or pitch of the roof varies with climate and aesthetic considerations and is expressed in inches fall per horizontal foot. A roof that slopes 5 inches (125 mm) for each foot (0.3 m) horizontally from the ridge would be described as a "five-in-twelve" roof (Figure 4.47). A roof designed to withstand a heavy snow load might have a twelve-in-twelve pitch (45-degree angle). Some church roofs, for example, are even steeper (Figure 4.48).

Pitched roofs are usually supported by wooden rafters, laminated beams, or wooden or steel trusses spanning the shortest distance between bearing walls. Where ceiling joists are used, they are fastened to the top plate and the rafters so that the entire assembly forms a series of triangles (Figure 4.49). In some cases, the ceiling may be omitted and the rafters become part of the interior decor (Figure 4.50). The supporting structure carries the pitched roof sheathing, which can be constructed of various materials such as plywood or planking. The sheathing is then covered by some sort of weather-resistant material (Figure 4.51).

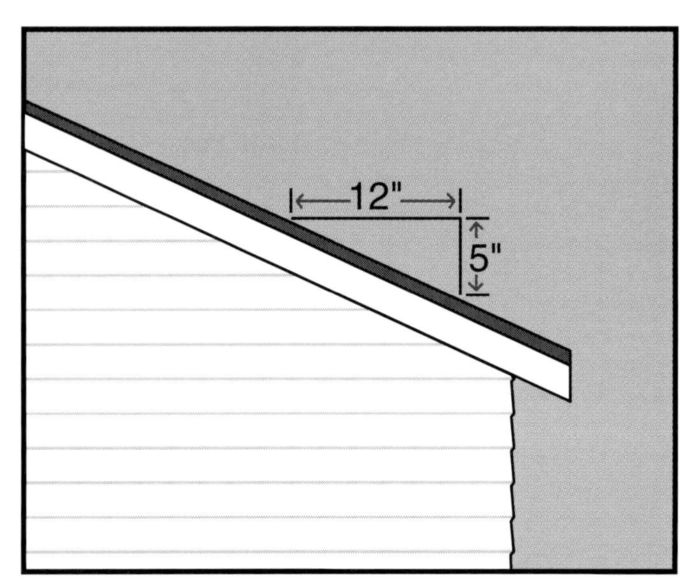

Figure 4.47 A five-in-twelve pitch.

Figure 4.48 A typical steep church roof.

In the average home, the space between the roof members and the ceiling forms an attic space, which may be vented by louvers at each gable end (Figure 4.52). Other types of buildings may have different roof vents (Figure 4.53). Access to attic spaces can usually be gained through a small attic scuttle in the

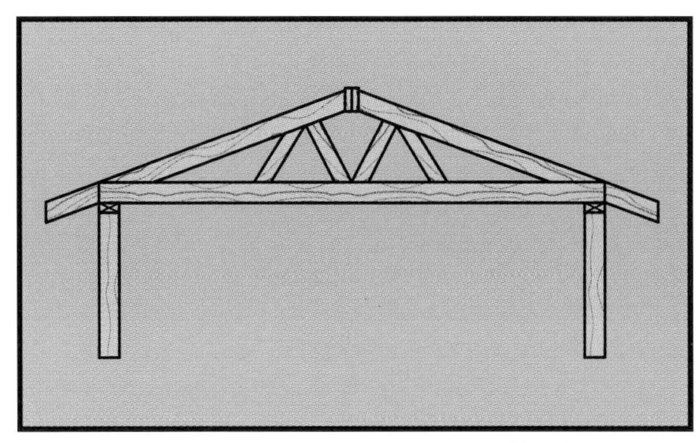

Figure 4.49 Triangles formed by typical pitched roof framing.

Figure 4.50 One form of open-beam ceiling.

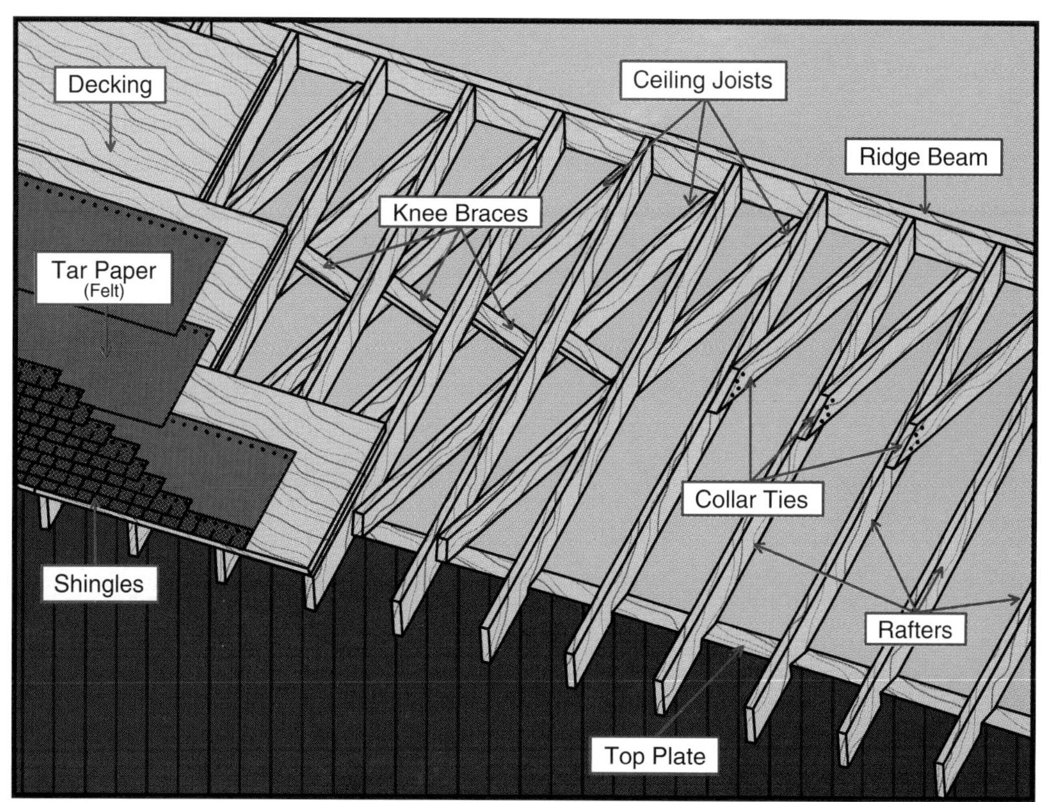

Figure 4.51 Components of various pitched roof systems.

Figure 4.52 A typical residential gable vent.

Figure 4.53 Eyebrow vents in a commercial building.

ceiling of a hallway or closet (Figure 4.54), or there may be a stairway or ladder leading into the space. Insulation material, which may or may not be combustible, is sometimes found between the framing under the roof or on top of the ceiling.

Typical roof coverings on pitched roofs are wooden shingles or heavier wooden shakes, composition shingles or rolled roofing, concrete or clay tiles, slate, or corrugated metal (Figures 4.55a-c).

Figure 4.54 A typical attic scuttle in the ceiling of a commercial building.

Figure 4.55a A typical composition-shingle roof.

Figure 4.55b A typical wooden-shingle roof.

Figure 4.55c A typical wooden-shake roof.

TYPES OF PITCHED ROOFS

Gable roof. The gable roof is perhaps the most common style of roof construction and can be found on most residential dwellings and many commercial structures (Figure 4.56). This type of roof is constructed in a semiflat to a very steep pitch, and the points where the rafters meet the outside walls and the ridge beam provide the most support. Its A-frame configuration consists of rafters that run perpendicular to the ridge beam and down to and usually beyond the outside walls. The ridge and rafters are often 2 x 6 inches or larger, and the rafters are commonly spaced at 16 to 24 inches (400 mm to 600 mm) on center. The size and spacing of the rafters will vary with the horizontal distance being spanned. Additional support may be provided by collar beams and ceiling joists. Valley rafters are used where two roof lines intersect (Figure 4.57). The trussed pitched roof is designed to cover a considerable span, and its rafters can be made of timber or metal (Figure 4.58).

Figure 4.56 A typical gable roof.

Hip roof. The hip roof is similar to the gabled roof in every respect except that the ends of the roof terminate in a "hip" configuration rather than a gable. In other words, the roof slopes down to meet every outside wall (Figure 4.59).

Figure 4.57 Typical valley rafter and pitched roof framing.

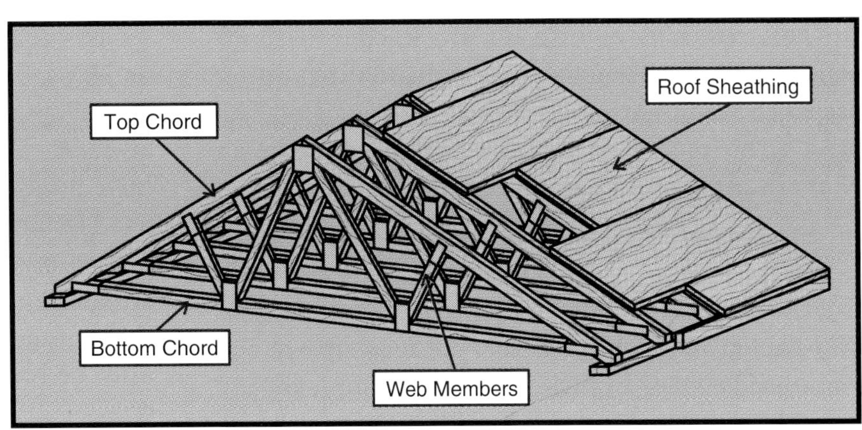

Figure 4.58 A typical trussed pitched roof.

Figure 4.59 A typical hip roof showing the absence of gables.

Hip roof construction consists of a ridge beam with conventional rafters running perpendicular to the ridge and hip rafters running from the end of the ridge beam, at a 45-degree angle, down to and beyond the outside walls at the corners (Figure 4.60). The rafters in the hip sections run in the same direction as the ridge beam. The dimensions and spacing of hip roof structural members follow the same engineering rules as those for gabled roof construction. The strongest parts of this roof system are the ridge beam, valley rafters, hip rafters, and the points where the rafters cross the outside walls.

Figure 4.60 Typical hip roof framing.

Lantern roof. The lantern roof consists of a high gabled roof with a vertical wall above a downward-pitched shed roof section on either side. This roof style is found on many barns, churches, and commercial buildings with rural-style construction (Figure 4.61).

Figure 4.61 A typical lantern roof.

This roof may be difficult to ventilate without an aerial device because of the difficulty in gaining access to the upper roof from either lower roof. The peak of the upper roof may also be beyond the reach of available ground ladders.

Shed roof. The shed roof can be seen as a flat roof sloped from only one side — usually from the front of the building down to the back — or as half of a gabled roof. This type of roof may utilize the mono-pitch truss (Figure 4.62), which employs only a single web member and therefore may be more prone to early collapse than other lightweight wooden trusses.

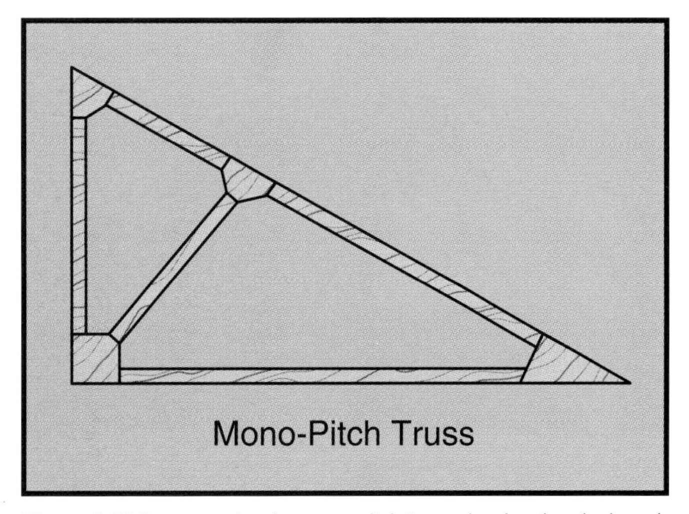

Figure 4.62 Components of a mono-pitch truss showing the single web member.

Bridge truss roof. *Bridge trusses*, also known as Howe or Pratt trusses, are heavy-duty trusses with sloping ends (Figure 4.63). When constructed of wood, the trusses are usually made from 2- x 12-inch lumber, and vertical metal tie rods may be used for additional support. Joists are usually 2 x 6 inches or 2 x 8 inches covered with 1- x 6-inch sheathing. Composition roofing may be used throughout this type of roof, or the sloping sections may be shingled. Because of the shape of bridge trusses, they form a roof that is very similar to the modern mansard roof discussed later in this section.

The strongest areas of bridge trussed roofs are at the perimeter of the building where the trusses and roof meet the outside bearing walls. The trusses are in constant tension and compression and will fail under severe fire conditions, but the likelihood of roof collapse is dependent on the dimensions of the materials used and the span of the trusses. If metal tie rods are used, early failure of the rods will also affect the stability of the trusses.

Mansard roof. The mansard roof has a double slope on each of its four sides. Instead of the roof running at a constant angle, there are two angles. One angle forms a steep pitch running from the eaves to a certain height, and the other produces a flatter pitch to the ridge of the roof.

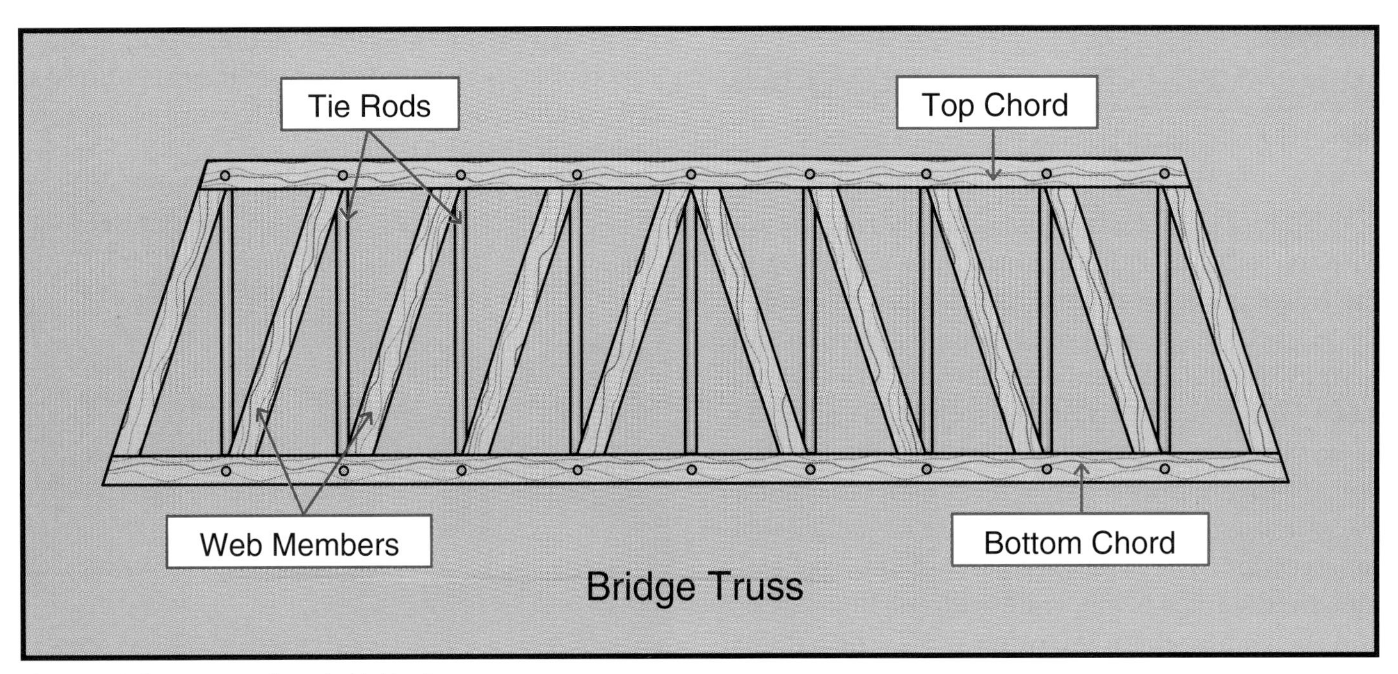

Figure 4.63 Components of a typical bridge truss.

This style is similar to the gambrel roof (see gambrel roofs later in this section) in that the lower slope is steeper than the upper slope. The difference between the mansard and the modern mansard (see following paragraph) is the way in which the four sides meet in the middle. The mansard forms a hipped peak or ridge (Figure 4.64), while the modern mansard forms a flat central portion.

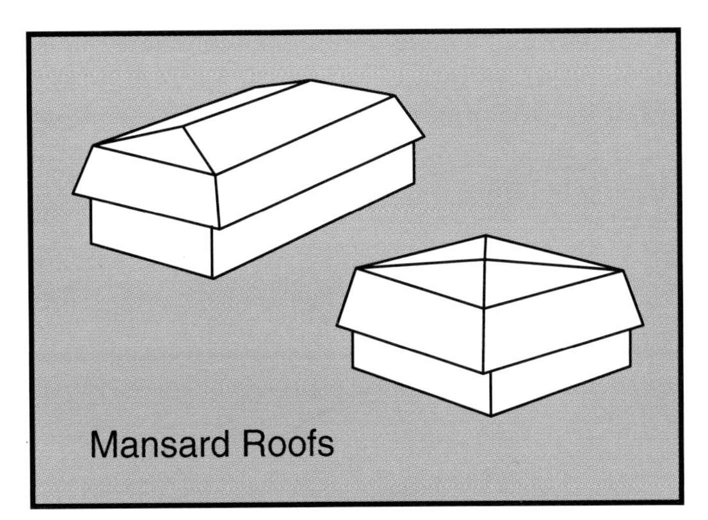

Figure 4.64 Two forms of the classic mansard roof.

Modern mansard roof. The modern mansard roof possesses characteristics of both flat and pitched roofs: four steeply sloped sides rise to meet a flat top called a *deck* (Figure 4.65). This roof type may utilize bridge trusses or K-trusses as supporting members, both of which allow for the creation of an ample void between the roof and the ceiling, as well as potential for early collapse under fire conditions. In addition, the modern mansard style roof includes overhangs that form concealed spaces through which fire and smoke can spread quickly (see modern mansard roofs under Types Of Flat Roofs section).

Gambrel roof. A gambrel roof is most often found on barns and other outbuildings. This roof is essentially a gable type with two different slopes on each side of the peak, with the lower slope being steeper than the upper slope (Figure 4.66). Such a design makes getting to the top of the roof very difficult with ground and roof ladders, so an aerial device may be the preferred method of access. Because this roof design permits efficient use of attic space under the roof, firefighters should consider that additional weight resulting from maximum interior attic storage could hasten the failure of roof supports during a fire.

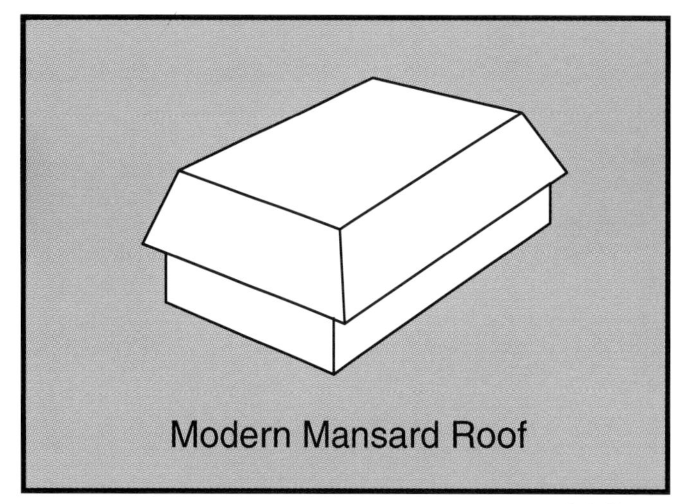

Figure 4.65 The modern mansard roof showing the characteristic flat top.

Figure 4.66 A typical gambrel roof.

Sawtooth roof. The sawtooth roof is used in industrial buildings to maximize light and ventilation. This design consists of a series of inclined planes similar in shape to the teeth of a saw (Figure 4.67). The rafters are 2 x 8 inches or larger and use wood or metal supports for bracing. The vertical walls include openable windows along their entire length, usually consisting of panes of wired glass. The sloped roof structures are sheathed with plywood or planked sheathing and covered with roofing material.

Butterfly roof. The butterfly roof may be seen as two opposing shed roofs that meet at their lower edges in the middle of the building (Figure 4.68). The same hazards and operational and safety con-

Figure 4.67 A typical sawtooth roof.

Figure 4.68 A butterfly roof showing the effects of a recent fire. *Courtesy of Wes Kitchel.*

siderations of all other pitched roofs also apply to butterfly roofs with one exception. A firefighter who slips and falls on a butterfly roof is not in jeopardy of sliding off the edge of the roof as might happen on other types of pitched roofs.

HAZARDS OF PITCHED ROOFS

Pitched roofs are designed to shed water and snow, and their major hazard may be the steepness of the roof and the lack of secure footing for firefighters working on them. This hazard is increased when the roof is wet or covered with ice, snow, wet leaves, etc.; but it may even be hazardous when dry because of the loose granular texture of some roof coverings. Loose roof tiles, slate, or broken pieces can also be hazardous to firefighters working on the roof and on the ground. Loose tiles and broken shards may cause firefighters on the roof to slip, and this debris can fall on firefighters working on the ground.

In addition to the hazards presented by the angle of the roof and by falling debris, the growing use of 2- x 4-inch trusses with metal gusset plates and ½-inch or ⅜-inch plywood decking can present an even greater hazard to firefighters. These assemblies offer little fire resistance and can fail rapidly during a fire, resulting in sudden roof collapse with little or no warning (Figure 4.69). As one way of mitigating this hazard, a growing number of fire departments have their dispatchers announce the elapsed time into an incident at 5, 10, 15, and 20 minutes. This practice helps the IC decide if and when interior crews should be ordered out of a burning building.

Figure 4.69 Typical lightweight pitched roof framing. *Courtesy of Wes Kitchel.*

VENTING PITCHED ROOFS

Pitched roofs should be vented at the highest point on the leeward side directly over the fire or as close to it as safely possible. The opening should be cut parallel to the rafters, perpendicular to the ridge, and to the size deemed necessary (Figure 4.70). In some cases, as with plywood roof decking, it may be advantageous to strip the roof covering before cutting and pulling the decking. In other cases, especially when using the center-rafter technique, leaving the roof covering attached works well and saves time (Figure 4.71). On metal-covered roofs it may be possible to remove an entire section at one time by cutting or prying along the edges, pulling screws or nails as necessary, and removing the panel.

Figure 4.70 A typical vent hole cut in a pitched roof.

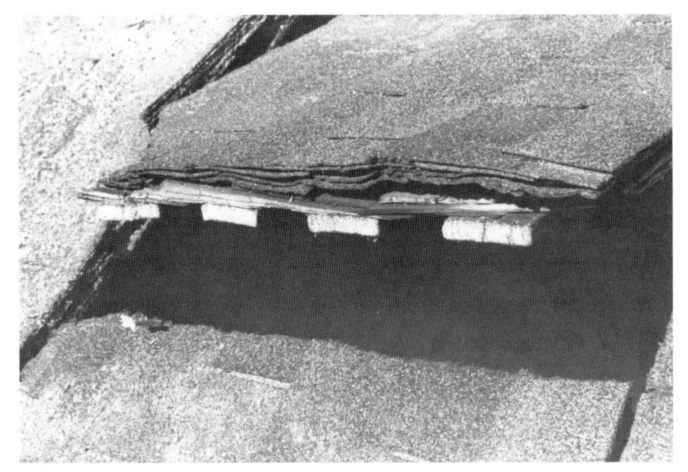

Figure 4.71 A louver vent with the roof covering left attached.

Because of the steep slope of many pitched roofs, the job of cutting a roof ventilation opening can be difficult and dangerous. Firefighters, preferably wearing ladder belts, should use roof ladders to prevent themselves from sliding down the roof (Figure 4.72). When working from a roof ladder, the firefighter will often need to reach as far as safely possible laterally from the ladder in order to cut the largest possible ventilation opening. To provide more secure footing, the pick of an axe can be embedded in the roof and the axe head used as a foothold (Figure 4.73). Long pike poles or rubbish hooks may also be needed when ventilating this type of roof because the ceiling can be several feet (meters) below the roof level, and sections of the ceiling may have to be removed for complete ventilation (Figure 4.74).

Flat Roofs

The general category of flat roofs includes the inverted roof, rain roof, wooden deck roof, metal deck roof, concrete roof, poured gypsum roof,

Figure 4.72 A firefighter working from a roof ladder to avoid sliding down the roof.

Figure 4.73 A firefighter using an axe head as a foothold.

Figure 4.74 A firefighter using the handle of a pike pole to push down the ceiling below.

modern mansard roof, and panelized roof. Flat roofs are more common on mercantile and industrial buildings, multiple dwellings, and in apartment complexes than on single-family dwellings. This type of roof ordinarily has a slight slope (two-in-twelve pitch or less) toward the rear of the building to permit drainage (Figure 4.75) and is frequently penetrated by chimneys, vent pipes, shafts, scuttles, bulkheads, and skylights. The roof may be surrounded and divided by parapets (Figure 4.76). It may also support water tanks, HVAC equipment, antennas, solar panels, billboards, and other dead loads that may interfere with ventilation operations and increase the likelihood of roof collapse (Figure 4.77).

Figure 4.75 A typical flat roof showing the slight slope.

Figure 4.76 A typical parapet wall dividing a large flat roof.

Figure 4.77 A dead load typical of those found on flat roofs.

Flat roofs are commonly supported by horizontal joists or rafters similar to the joists used in floor systems. The structural elements of flat roofs consist of a wooden, concrete, or metal substructure covered with sheathing (Figure 4.78). The sheathing is, in some cases, covered with a layer of dense foam insulation under the weatherproof finish layer. There is usually a concealed, and possibly unvented, space between the flat roof and the ceiling of the top floor below. This space is referred to as an *attic, cockloft, crawl space*, or *interstitial space*. The often unprotected underside of the roof assembly will be exposed to the effects of any fire in this concealed void, which may contribute to early roof failure.

TYPES OF FLAT ROOFS

***Inverted roof*.** Inverted roofs differ from conventional flat roofs primarily in the location of their main roof beams. In a conventional roof system the main joists are set at the final roof level, sheathing is nailed to the tops of the joists, and a ceiling may be nailed to the bottoms of the joists, or more commonly, it will be suspended below the joists, creating a concealed space. In the inverted roof, the main roof joists are set at the level of the ceiling, and a framework of 2- x 4-inch members is constructed above the main joists. The sheathing is nailed to this framework, and the weatherproof covering is applied over the sheathing. This roof is a fairly solid roof system that retains its structural integrity until the upright members burn through.

From the outside, the inverted roof looks like any other flat roof, but the surface usually feels "springy" or "spongy" to anyone walking on it. The

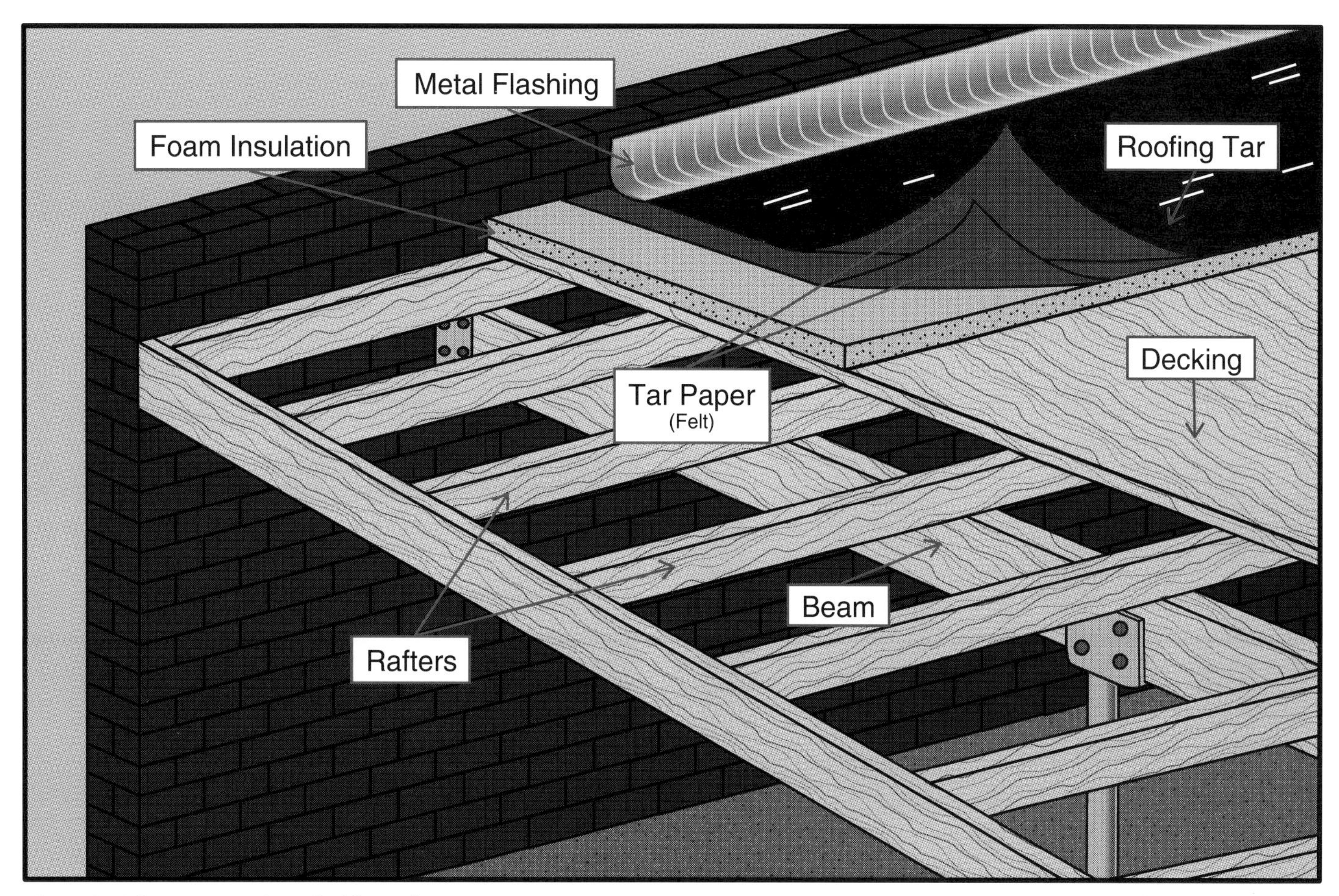

Figure 4.78 Components of a typical flat roof.

design of the inverted roof creates a concealed space several feet in height between the ceiling and the roof deck. The unprotected structural members within this concealed space are exposed on all four sides, so they are subject to severe damage from fire within the space. Crews should familiarize themselves with any of these roofs in their districts by making thorough pre-incident planning inspections (Figure 4.79).

Figure 4.79 Firefighters testing an inverted roof for sponginess during a pre-incident planning inspection.

Rain roof. One dangerous relative of the inverted roof is the so-called "rain roof." While this oddity can be found over any type of roof, it is most common on buildings with flat or arched roofs. The rain roof is built *over* an existing roof that has become so porous it does not keep out the rain and/or it sags sufficiently to allow rainwater to collect on the roof. The existing roof is left in place, and the new roof is built on a raised framework above the original (Figure 4.80).

Figure 4.80 The badly deteriorated original roof shows through the vent hole cut in this rain roof.

This practice creates some potential problems. First, the void created between the two roofs may allow fire to burn undetected for some time and result in an inaccurate size-up of the fire. Second, the existence of two separate roofs can seriously impede effective ventilation or prevent it entirely. Finally, the original roof was not designed to support the additional weight, and therefore the entire roof assembly may be more susceptible to collapse.

***Wooden deck roof*.** Wooden deck roofs may present a hazard if lightweight plywood decking is used. Plywood of ⅜- to ⅝-inch thickness offers little fire resistance, and it may be difficult to remove for ventilation purposes (Figure 4.81).

Figure 4.81 The underside of a typical lightweight wooden deck roof. *Courtesy of Wes Kitchel.*

CAUTION: Some brands of fire-resistant plywood have been known to delaminate and weaken over time, resulting in a deck that a firefighter can step through without it having suffered any fire damage. Identifying these hazardous roofs during pre-incident inspections is critical to firefighter safety.

Roofs sheathed with wooden planks are easier to strip but may be somewhat less wieldy when making louver vents or trench cuts (Figure 4.82). The structural stability of the joists will vary depending on the span, the size and spacing of the joists, and whether the joists are suspended by metal hangers (Figure 4.83).

***Metal deck roof*.** Metal deck roofs consist of metal bar joists, which usually run across the narrow dimension of the building, and metal decking that is laid perpendicular to the joists (Figure 4.84). In most cases, the metal decking is spot welded to the joists. Large-area metal deck roofs consist of large supporting beams that run across the narrow dimension of the building and bar joists that run perpendicular to the beams and parallel to the long dimension of the building (Figure 4.85).

Figure 4.82 The underside of a wooden deck roof sheathed with planks.

Figure 4.83 Heavy roof joists suspended by metal hangers.

CAUTION: Metal deck roofs can be expected to fail within a very few minutes of flame impingement. Because the heat of a fire will soften metal and make it more pliable, metal decking around roof vents and other openings may not support the weight of a firefighter.

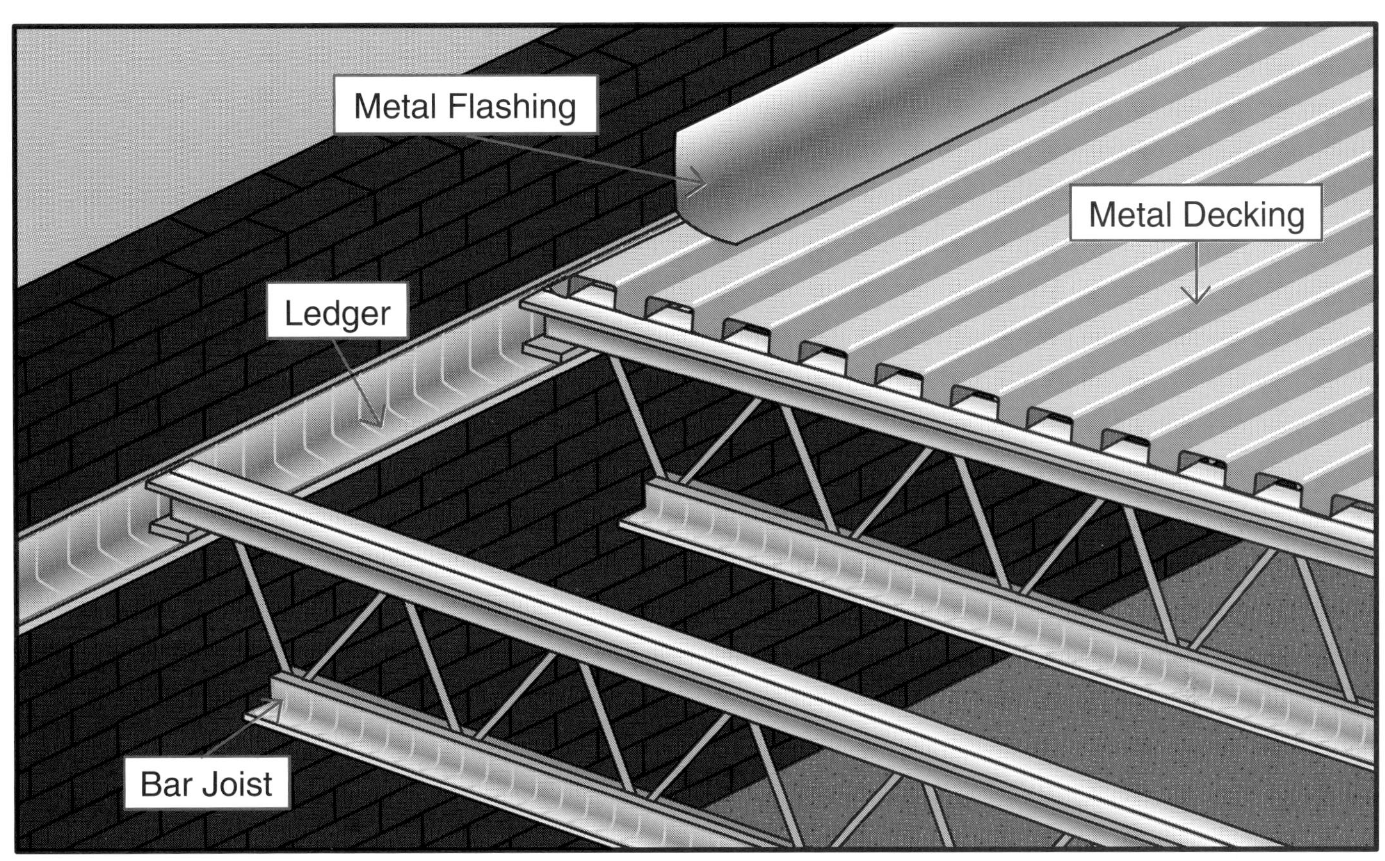

Figure 4.84 Components of a typical metal deck roof.

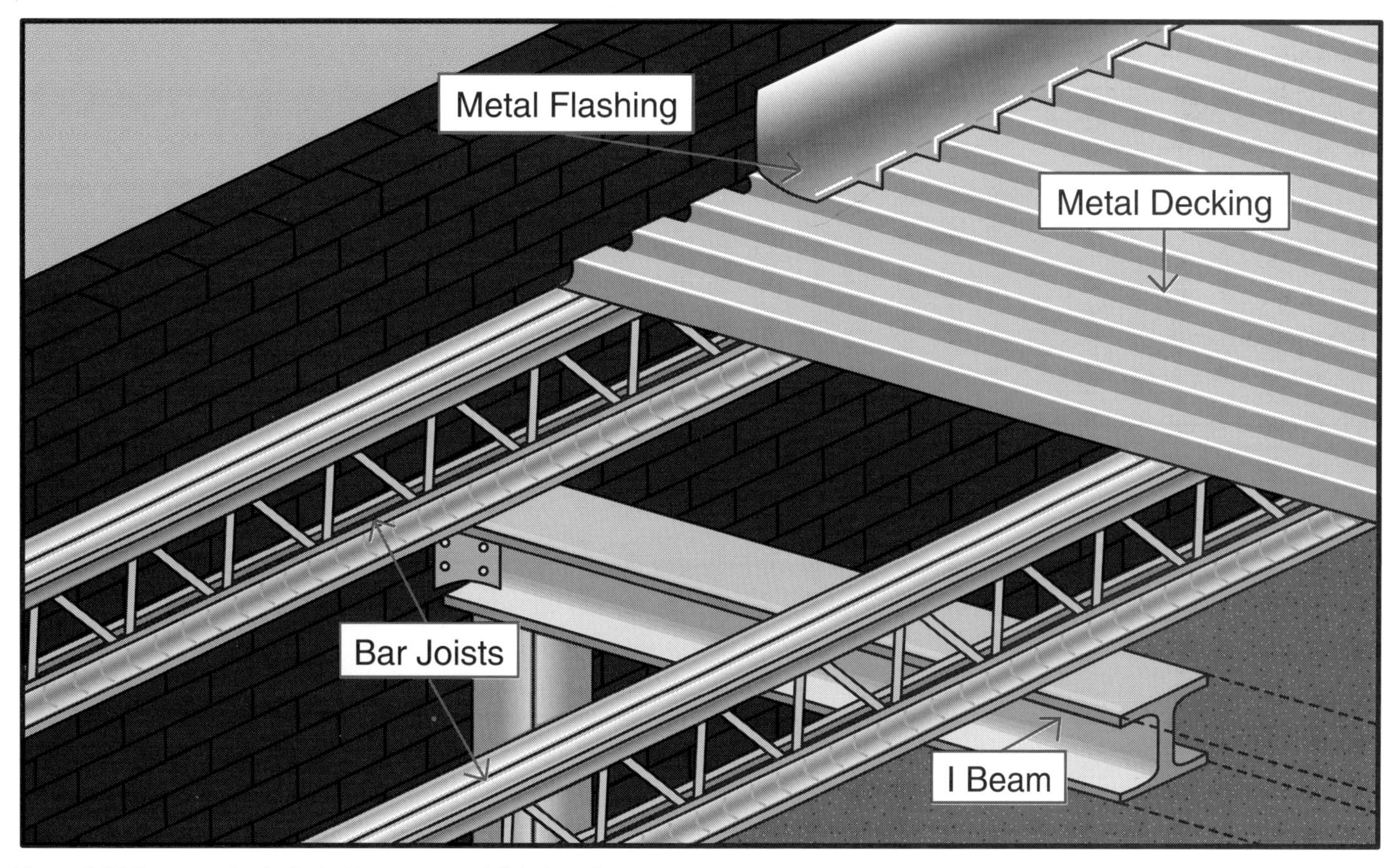

Figure 4.85 Components of a typical large-area metal deck roof.

***Concrete roof*.** Concrete roofs are constructed in a variety of ways. The two most common are made of lightweight concrete poured over metal decking (Figure 4.86) or of precast "Double-T" panels (Figure 4.87). This type of roof provides a smooth, hard surface that is structurally strong and highly resistant to fire, but it may be extremely difficult to breach for ventilation purposes. Opening these roofs may require special tools, such as jackhammers, core drills, or burning bars, and will be a laborious and time-consuming operation.

Lightweight concrete roofs may be cut using a rotary saw with a masonry blade. Because it is difficult and time consuming to cut concrete roofs that are over 4 inches (100 mm) in thickness, they are often designed with built-in access panels that can be lifted out in an emergency. Using existing openings, such as bulkheads, ventilators, or scuttles, will certainly be the fastest way to create an opening in these roofs and may be the only practical way. Once again, thorough pre-incident familiarization will greatly speed the process of ventilating these roofs.

***Poured gypsum roof*.** Poured gypsum roofs consist of bar joists or I beams with bulb tees tack welded to the joists. Gypsum board is placed upon the bulb tees and is covered with a layer of gypsum cement up to 2½ inches (65 mm) thick to which wire mesh reinforcement is added. The gypsum is then sealed with a weatherproof covering.

Because these roofs are constructed of materials that are highly resistant to fire, they retain their structural stability longer than some other roof types. The roof covering is easily cut with a power saw using a metal-cutting blade, and the covering can then be rolled back to open the hole.

***Modern mansard roof*.** This type of roof has characteristics of both pitched and flat roofs. The perimeter of the roof consists of steeply pitched sections that surround a flat roof area in the middle (Figure 4.88). These roofs are most commonly

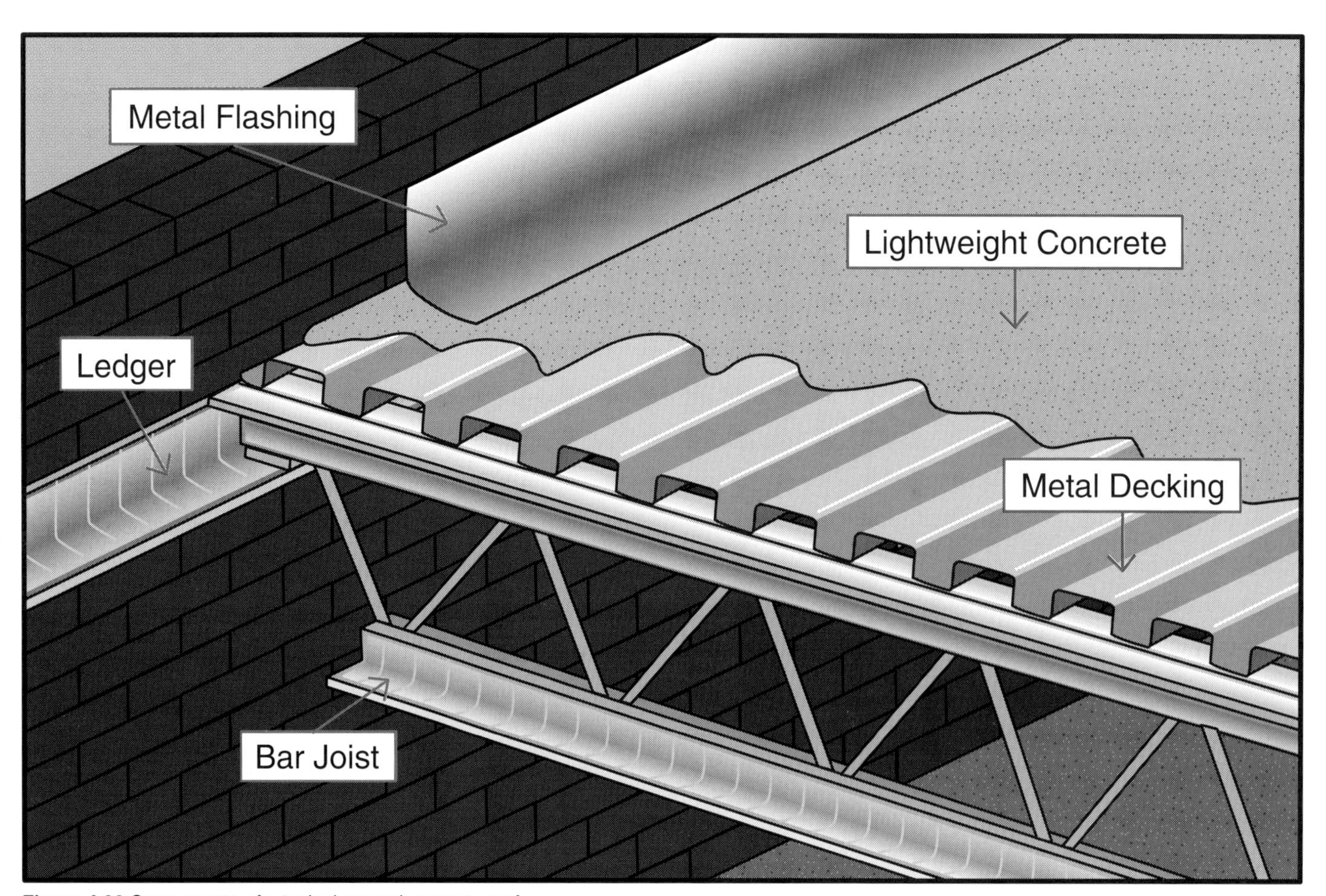

Figure 4.86 Components of a typical poured concrete roof.

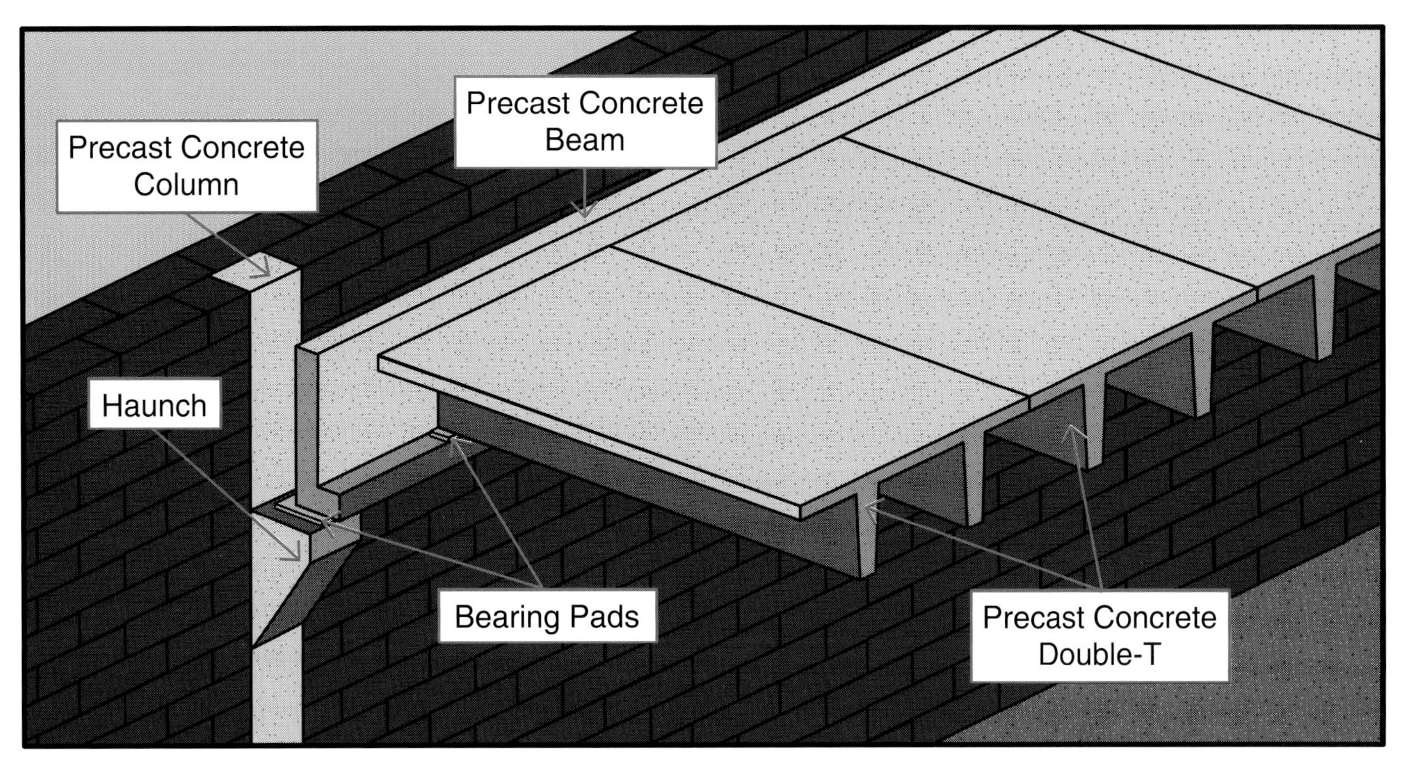

Figure 4.87 A concrete roof made up of precast concrete Double-T panels.

Figure 4.88 A typical modern mansard roof.

supported by bridge trusses, and the same operational and safety considerations apply to these roofs as to any other pitched or flat roof, with a couple of exceptions.

First, some modern mansard roofs are actually only false roofs, each consisting of a wall (with a triangular cross section) that has been added to the perimeter of a flat roof for aesthetic reasons. This construction creates a depressed area behind the false roof that can range from a few feet to several feet deep. Unsuspecting firefighters can fall from the top of this false roof onto the actual flat roof if visibility is obscured by smoke and/or darkness.

Second, in building these false roof structures, an uninterrupted concealed space that may run the entire perimeter of the roof is often created. This arrangement can allow fire to travel around the entire roof undetected. These false roofs usually extend beyond the exterior wall, creating an overhang that can collapse if fire weakens the bracing on the original roof (Figure 4.89).

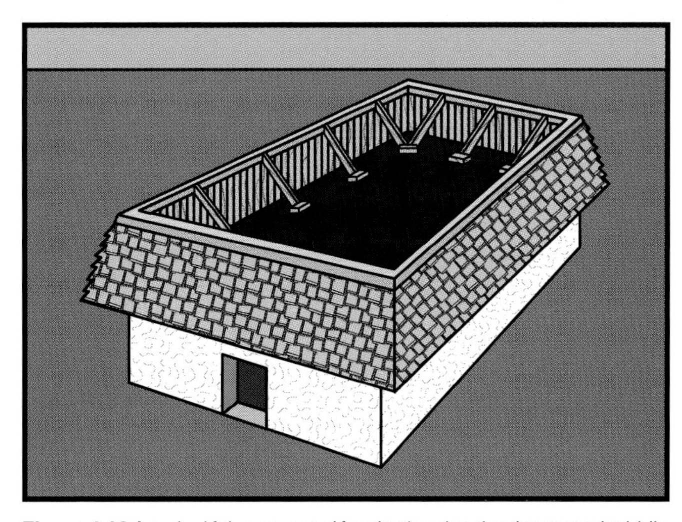

Figure 4.89 A typical false mansard fascia showing the depressed middle area and typical bracing.

***Panelized roof*.** Many modern buildings have panelized flat roofs. These increasingly common roofs are discussed later in this chapter under the Lightweight Construction section.

HAZARDS OF FLAT ROOFS

Firefighters preparing to ventilate a flat roof should look for hot spots or sagging of the roof area *prior* to walking on the roof surface and should look *continually* for these signs during ventilation operations. The roof should always be sounded before stepping onto it and repeatedly sounded when firefighters are moving about on the roof. Hot spots may be recognized by melting, soft, or bubbling tar; by melting snow and ice; and by waves of heat rising from one area. Sagging indicates damage to the substructure of the roof assembly. Each of these signs indicates severe heat and fire conditions directly below the roof and suggests that the roof may be ready to collapse, either partially or totally. Inspection holes can be cut to help firefighters monitor fire conditions (Figure 4.90). Crews working on flat roofs must exercise extreme caution if these signs exist or develop while work is in progress.

Figure 4.90 Firefighter cutting an inspection hole to monitor fire conditions below.

Some older buildings with flat roofs may have wooden access ladders built onto the side of the structure (Figure 4.91). These ladders should generally not be used by firefighters because age and weathering may have made them unsafe. If access from adjacent roofs is not possible, fire department ground ladders or aerial devices should be used to gain roof access for ventilation.

Overhangs are often added to flat roofed buildings to give the appearance of a mansard roof (Figure 4.92). These overhangs form concealed spaces through which fire and smoke can quickly spread undetected.

Inverted flat roofs create special hazards as well. This design also creates a concealed space, often several feet (meters) in height, which includes many unprotected wooden structural members. Heavy fire conditions can burn quickly through the 2- x 4-inch supporting members, causing the roof deck to collapse onto the roof joists.

Other hazards encountered on flat roofs are the security measures building owners have taken to deter burglars from entering through skylights and other roof openings. Some have installed barbed wire or razor ribbon around the perimeter of the roof (Figure 4.93). Guard dogs can also be found on flat roofs of business and apartment buildings.

Figure 4.91 An older building with a built-in wooden access ladder.

Figure 4.92 A typical overhang created by adding a false mansard fascia.

Figure 4.93 Razor ribbon protecting the roof of a commercial building. *Courtesy of Bob Ramirez.*

These dogs can delay or prevent access to a roof, so their presence should be determined during pre-incident inspections. All of these security measures can injure firefighters and/or delay access to the roof for ventilation.

Many buildings with flat roofs also have parapet walls that can be both a hazard and a help to firefighters during ventilation operations. These walls may extend from a few inches to several feet above the roof's surface. High parapet walls create a potential fall hazard, and walls that are too low may cause firefighters to trip and fall over them (Figure 4.94). Before stepping off a parapet wall, especially if the roof is obscured by smoke or darkness, firefighters should always sound the roof by striking it with the blunt end of an axe or other tool. This process will reveal the condition of the roof as well as the vertical distance from the top of the wall to the roof. Properly constructed parapet walls can help prevent the spread of fire from building to building and can help prevent firefighters from accidentally falling or walking off the roof. Because the heights of parapet walls vary so much from building to building, firefighters must become familiar with those in their response districts through pre-incident inspections.

Figure 4.94 A firefighter inspecting a low parapet wall on a flat roof.

VENTING FLAT ROOFS

As with pitched roofs, flat roofs should be vented as close to directly over the seat of the fire as is safely possible. Flat roofs may be vented by removing roof vents, cutting a large square or rectangular ventilation hole, cutting a strip or trench ventilation opening, or using some combination of these methods.

When smoke under pressure is issuing from a roof ventilator, the ventilator should be removed to facilitate the movement of the smoke and fire gases (Figure 4.95). In some cases, it will also be advantageous to enlarge the opening by cutting the metal shroud down to the flashing and folding the shroud back (Figure 4.96). At least two sides of roof monitors should be removed.

Figure 4.95 A roof ventilator removed to facilitate the movement of smoke and fire gases.

Figure 4.96 The ventilator shroud cut down to open it up further.

If bubble skylights need to be opened for ventilation, they should be removed from their frames (Figure 4.97), and the panes in wired glass skylights should be removed or broken. Before breaking glass in skylights, interior crews should be warned. Firefighters should break a single pane first and then pause before breaking out the remaining panes. The pause will give firefighters who did not hear the warning an opportunity to move out of the way or to alert the vent crew of the need to delay breaking the remainder of the panes.

Figure 4.97 Firefighters removing bubble skylights on a commercial building.

To create sufficiently large exit openings for efficient ventilation of buildings with flat roofs, cutting ventilation holes may also be necessary. Many flat roofs have a thick covering of tar and gravel or other material that will need to be removed or cut before cutting the sheathing. An axe can be used to cut the roof covering or to scrape away some of the gravel to facilitate cutting with a power saw (Figure 4.98). Thick tar coverings tend to gum up chain saws, so a rotary saw may be a better choice.

Figure 4.98 A firefighter scrapes away gravel where saw cuts will be made.

Ventilation holes in flat roofs should be cut parallel to the rafters and perpendicular to the side walls of the building. After marking the perimeter of the opening to be cut, and working with the wind at their backs, firefighters should start by making a cut across the leeward end (Figure 4.99). This should be followed by making parallel side cuts approximately 4 feet (1.2 m) apart between the rafters. These cuts should start at the ends of the first cut and be made as long as necessary to create an exit opening of the required size. The roof covering and the sheathing may then be pulled back, as a unit or separately, with two pick-head axes, pike poles, or rubbish hooks (Figure 4.100). If it is necessary to break out a section of ceiling below the exit opening, a rubbish hook or the butt of a pike pole works well.

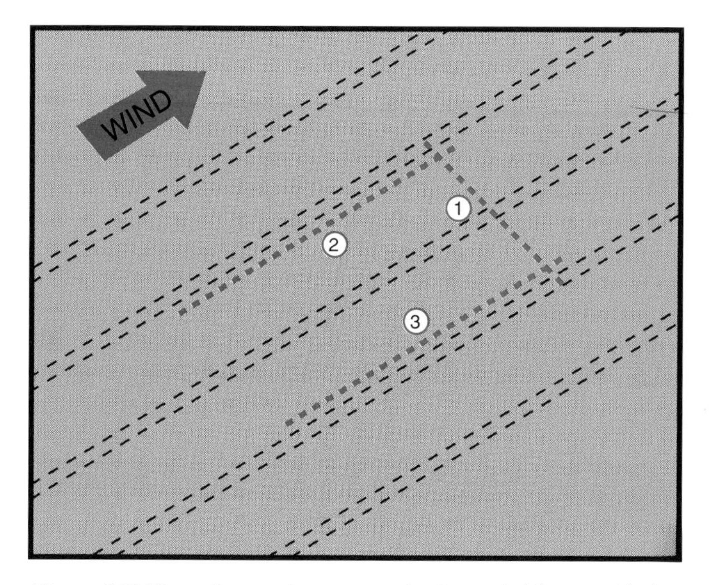

Figure 4.99 The pattern and sequence of cuts needed for a vent hole.

Center rafter or louver vents may be cut in flat roofs in the same manner as described previously, but with the addition of a fourth cut connecting the other ends of the side cuts. If it is necessary to cut a strip or trench vent, the louver vents may simply be extended from one outer wall to the other across the full width of the building (Figure 4.101).

Arched Roofs

Arched construction is typically used to support roofs with large, clear spaces such as exhibition halls, sports arenas, and similar structures. The most common types are ribbed (trussed), bowstring, and lamella arches.

Arched roofs may be supported with ribbed trusses, bowstring trusses, or arches of steel, concrete, or laminated wood. Steel arches can be made

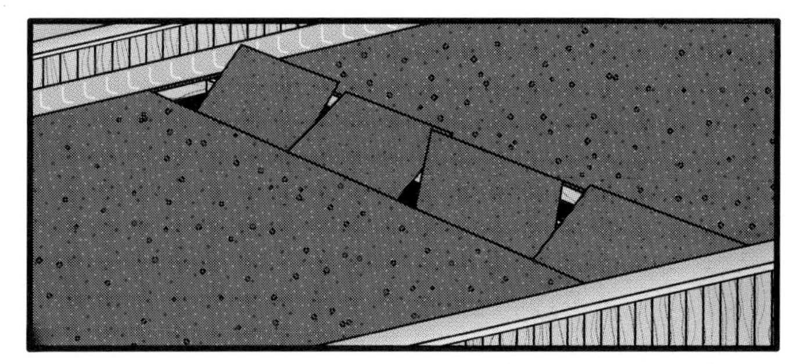

Figure 4.101 Louver vents extending from one outer wall to the other to form a trench vent.

Figure 4.100 Firefighters pulling the roof covering to expose the roof deck.

from plate girders or trusses (Figure 4.102). Wooden arches are laminated and glued in the factory.

Horizontal as well as vertical forces continuously act on an arched roof. As gravity attempts to flatten the arch, vertical forces bear down on the top of the arch. The resulting horizontal forces act on the ends of the arch, attempting to force them apart. These forces are resisted by abutments or buttresses at the ends of the arch (Figure 4.103) or by tension cables or tie rods between the ends of the arch (Figure 4.104). While the buttresses are usually stable, tension cables will lose their strength and integrity if exposed to fire. Arches may also contain hinges to permit rotation and thermal expansion. These hinges can be found at the top of the arch or at the abutments.

Figure 4.102 The underside of a typical steel truss arched roof.

Figure 4.103 Typical concrete buttresses at the ends of wooden arches.

Figure 4.104 Tension cables in an arched roof.

TYPES OF ARCHED ROOFS

Ribbed arch roof. The ribbed arch roof is similar in construction to the bridge truss roof, but its top chord is curved instead of straight (Figure 4.105). This roof can be found on older commercial-type structures and is usually made of 2- x 12-inch or larger wooden members.

Figure 4.105 The underside of a ribbed arch roof.

Bowstring arch roof. The bowstring arch roof, commonly found in bowling centers and supermarkets, uses tie rods for lateral support and turnbuckles to maintain proper tension (Figure 4.106). The main supporting members are easy to locate from the outside if the tie rods pass through the exterior wall to an plate or reinforcement star (Figure 4.107). The chords of these arch members are usually laminated 2- x 12-inch or larger lumber. The rafters (usually 2 x 10 inches) are covered by 1- x 6-inch sheathing and composition roofing material. The perimeter of the building and the arch members are the strongest points.

This roof is normally quite strong because of the size of the lumber used in its construction; however, bowstring arches have a history of early and sudden collapse when the tie rods are exposed to fire. Because the tie rods act to hold the outer walls together, the walls can be pushed outward when the tie rods fail, causing the entire building to collapse (Figure 4.108).

WARNING

Because of the potential for sudden collapse associated with bowstring arch roofs, incident commanders must be extremely cautious about placing crews inside these buildings, and must carefully monitor elapsed time when deciding when to withdraw them.

Figure 4.106 Tie rods in a bowstring arch roof. *Courtesy of Wes Kitchel.*

Figure 4.107 Reinforcement stars on the outside of an older building.

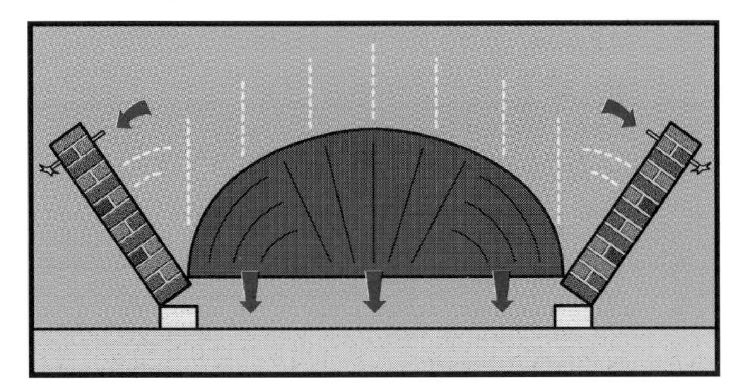

Figure 4.108 The results of tie rods being exposed to fire.

Lamella roof. The lamella roof is made up of a geometric, egg-crate or diamond pattern framework on which plank sheathing is laid. The framework is constructed of 2- x 12-inch wooden members bolted together at the intersections with steel gusset plates. The roof decking is 1- x 6-inch planking covered with composition roofing material. Support is provided by exterior buttresses or internal tie rods and turnbuckles. The perimeter of the building is the strongest area. The lamella roof system shares many characteristics with the bowstring-type roof, and the same operational and safety considerations apply.

HAZARDS OF ARCHED ROOFS

Firefighters can estimate the hazards of arched roofs by the size of the lumber and the span of the arches. In trussed arch roofs, the lower chord of the truss may be covered with a ceiling to form an enclosed cockloft or attic space. These concealed spaces are a definite impediment to effective ventilation and contribute to the spread of fire.

The single biggest hazard of arched roofs, however, is the danger of sudden and total collapse, often without warning (Figure 4.109). Because of this potential and because the rounded surface makes the use of roof ladders difficult, it is recommended that personnel work only from aerial devices when ventilating arched roofs (Figure 4.110).

WARNING

Heavy fire involvement of the truss area should prompt the immediate withdrawal of all personnel from the roof and from the interior of the building.

Figure 4.109 A collapsed arched roof. *Courtesy of Harvey Eisner.*

Figure 4.110 A firefighter working from an aerial device to vent an arched roof.

VENTING ARCHED ROOFS

Arched roofs should be ventilated at the top of the arch directly over the fire or by a long, narrow strip vent along the centerline of the roof. A conventional square opening can be cut perpendicular and adjacent to a main arch support (Figure 4.111), or a louver vent (which will probably be faster) can be made. If a strip vent is to be cut along the centerline of the roof, a series of louver vents may be the best choice (Figure 4.112).

Figure 4.111 A conventional square ventilation opening cut in an arched roof.

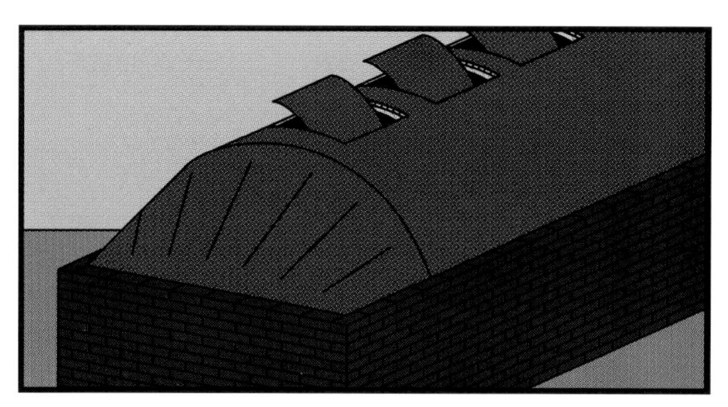

Figure 4.112 A series of louver vents cut along the centerline of an arched roof.

LIGHTWEIGHT CONSTRUCTION

Because of the continued escalation in the cost of labor and building materials, lightweight building and roof construction has become much more common. In modern construction, heavy timber, laminated beams, and 1- x 6-inch sheathing have given way to 2- x 4-inch lumber and ½-inch plywood, regardless of building size. Because these lightweight materials are less fire resistive than traditional materials, firefighters have less time in which to ventilate before the roof becomes unstable. This section focuses on the four major types of lightweight roof construction: panelized, open web truss, metal gusset plate truss, and wooden I beam.

Panelized Roofs

Panelized roof construction consists of laminated beams of various sizes (commonly 6 x 36 inches) that span the length or width of the building. These beams are supported at their ends by pilasters, wooden or steel posts, or saddles. Additional wooden or steel posts may provide support at intervals along the span. The beams may be bolted together to form lengths well in excess of 100 feet (30 m) and may be spaced between 12 feet (3.65 m) and 40 feet (12.3 m) apart. Wooden purlins (usually 4 x 12 inches) with metal hangers are installed on 8-foot (2.4 m) centers between and perpendicular to the beams (Figure 4.113). Wooden joists (usually 2 inches x 4 inches x 8 feet) are installed with metal hangers on 2-foot (0.6 m) centers between and perpendicular to the purlins (Figure 4.114). Sheets of plywood (4 feet x 8 feet x ½ inch) are nailed over this framework and then covered with composition roofing material. The strongest parts of this construction are the beams, the purlins, and the perimeter of the building where the roof meets the exterior walls.

A three-layer, laminated insulation paper is used on the underside of panelized roof decking. This material offers little protection to the joists and plywood decking because it consists of a tar-impregnated layer covered on either side by a layer of thin aluminum foil. When the insulation paper is subjected to fire, the foil peels away from the tar-impregnated paper and disintegrates, allowing the joists and decking to be exposed to fire.

Figure 4.113 Basic structure of a panelized roof showing the beams and purlins.

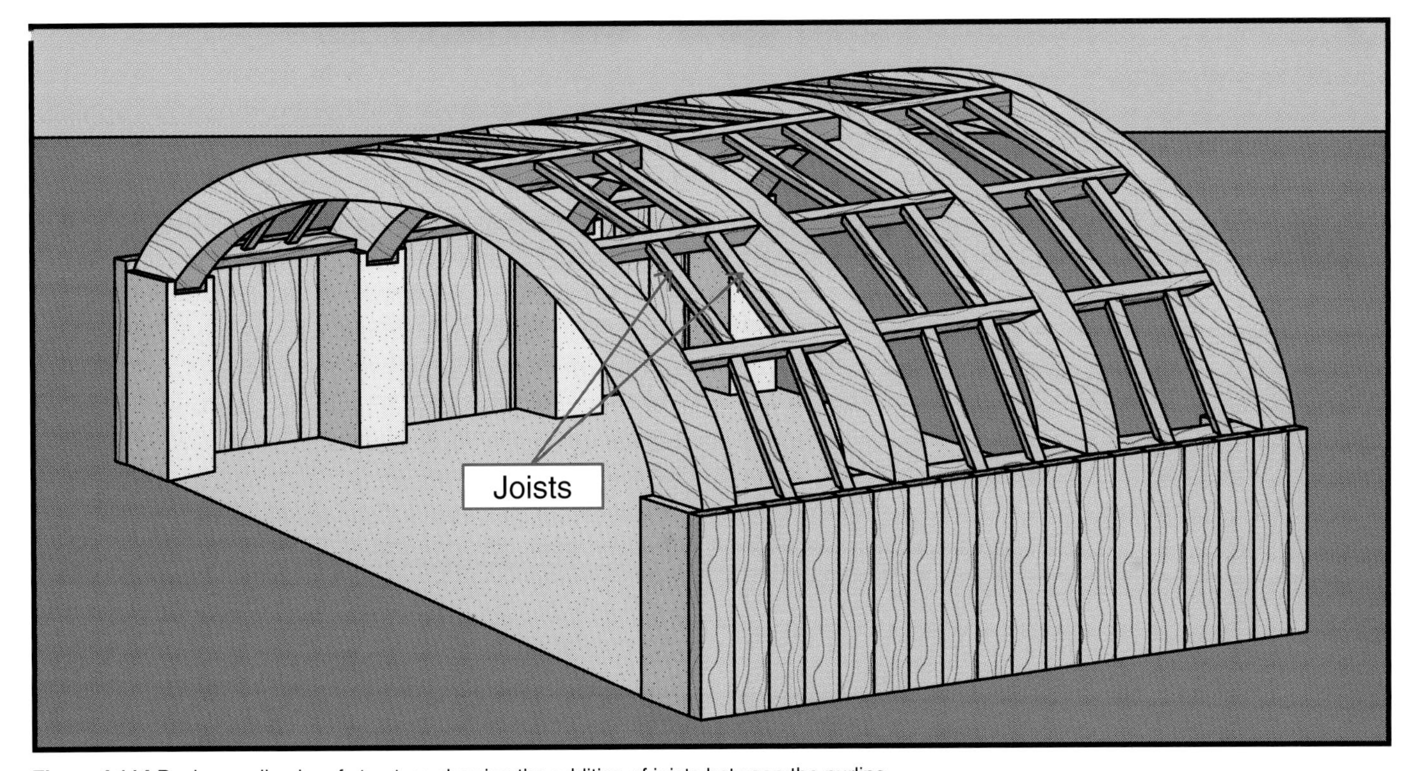

Figure 4.114 Basic panelized roof structure showing the addition of joists between the purlins.

Open Web Trusses

Open web trusses are prefabricated systems that consist of wooden top and bottom chords that are cross-connected by steel tube web members (Figure 4.115). The bridging effect of this system causes the top chord member to be in compression and the bottom chord to be in tension. The steel tube web members are made from 1- to 2-inch cold-rolled steel tubing with the ends pressed flat into a semicircular shape and a hole punched through each end. These flattened ends are then inserted into slots in the chords, and steel pins are driven through the flattened ends of the chord members, completing the assembly (Figure 4.116). The open configuration of these trusses creates large, open spaces through which fire can spread rapidly.

Figure 4.115 Open web trusses in position during building construction.

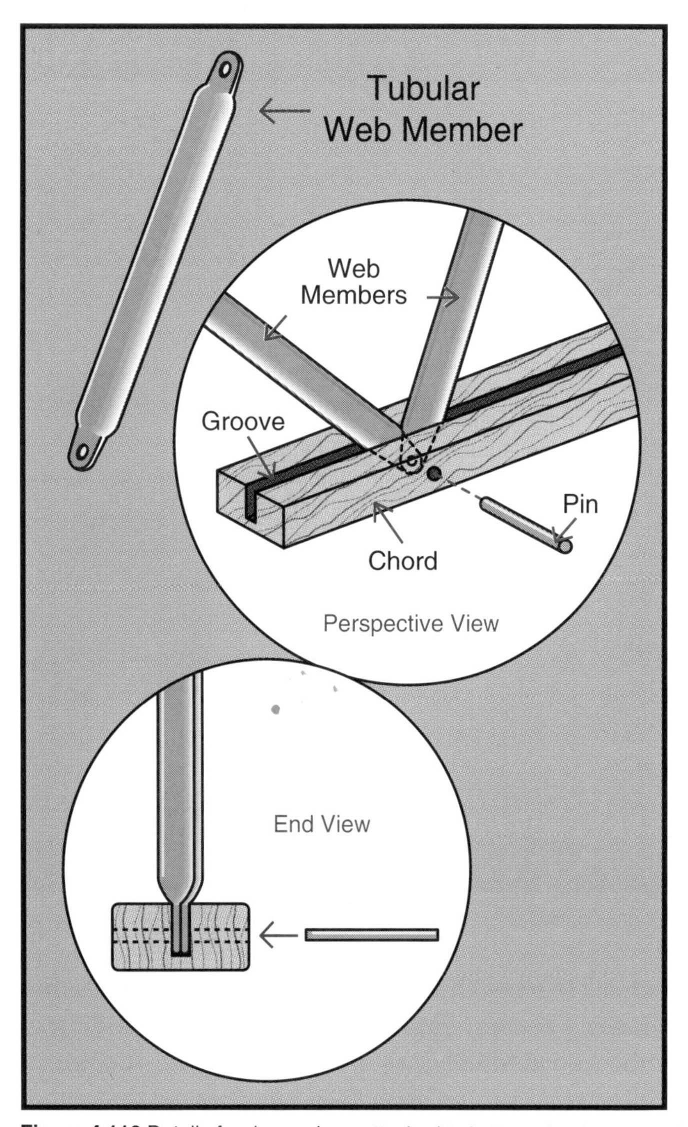

Figure 4.116 Detail of web members attached to bottom chord.

In a normal installation, the top chord rests on and is supported by a bearing wall, and the bottom chord is unsupported (Figure 4.117). Spans of up to 70 feet (21.5 m) are attainable using single 2- x 4-inch (or two 2- x 3-inch) members as top and bottom chords. By joining them with mitered and glued "finger joints," continuous 2- x 4-inch members exceeding lengths of 20 feet (6.2 m) are possible (Figure 4.118). Normal spacing of the joists is 2 feet (0.6 m) on center, and the area where the roof meets the exterior wall is the strongest point.

WARNING

Case studies have suggested that one reason for the sudden collapse of this type of roof is that interior crews inadvertently pull on the bottom chord of open web trusses when pulling ceilings.

Metal Gusset Plate Trusses

In this type of roof system, wooden trusses, used mostly in residential and light commercial applications, are usually made up of 2- x 4-inch lumber held together by metal gusset plate connectors (also known as *gang nails*) where the members intersect (Figure 4.119). The gusset plates vary in size, thickness, and depth of penetration, but plates with ⅜-inch prongs are most common. In trusses made up of 2- x 4-inch lumber, spans of up to 55 feet (17 m) are possible. The most common spacing between trusses is 2 feet (0.6 m) on center, and ½-inch plywood is commonly used as sheathing. The area where the trusses cross the outside bearing walls is the strongest point.

Trusses are normally supported only by the outside bearing walls (Figure 4.120). In this type of construction, interior partition walls are essentially freestanding walls that do not actually support the truss at any point. However, truss clips may be nailed to the bottom chord of the truss and to the top plate of a wall where the chord crosses a partition wall to provide lateral support for the wall.

Unless they are also corner-nailed, gusset plates often distort and pull out when exposed to fire.

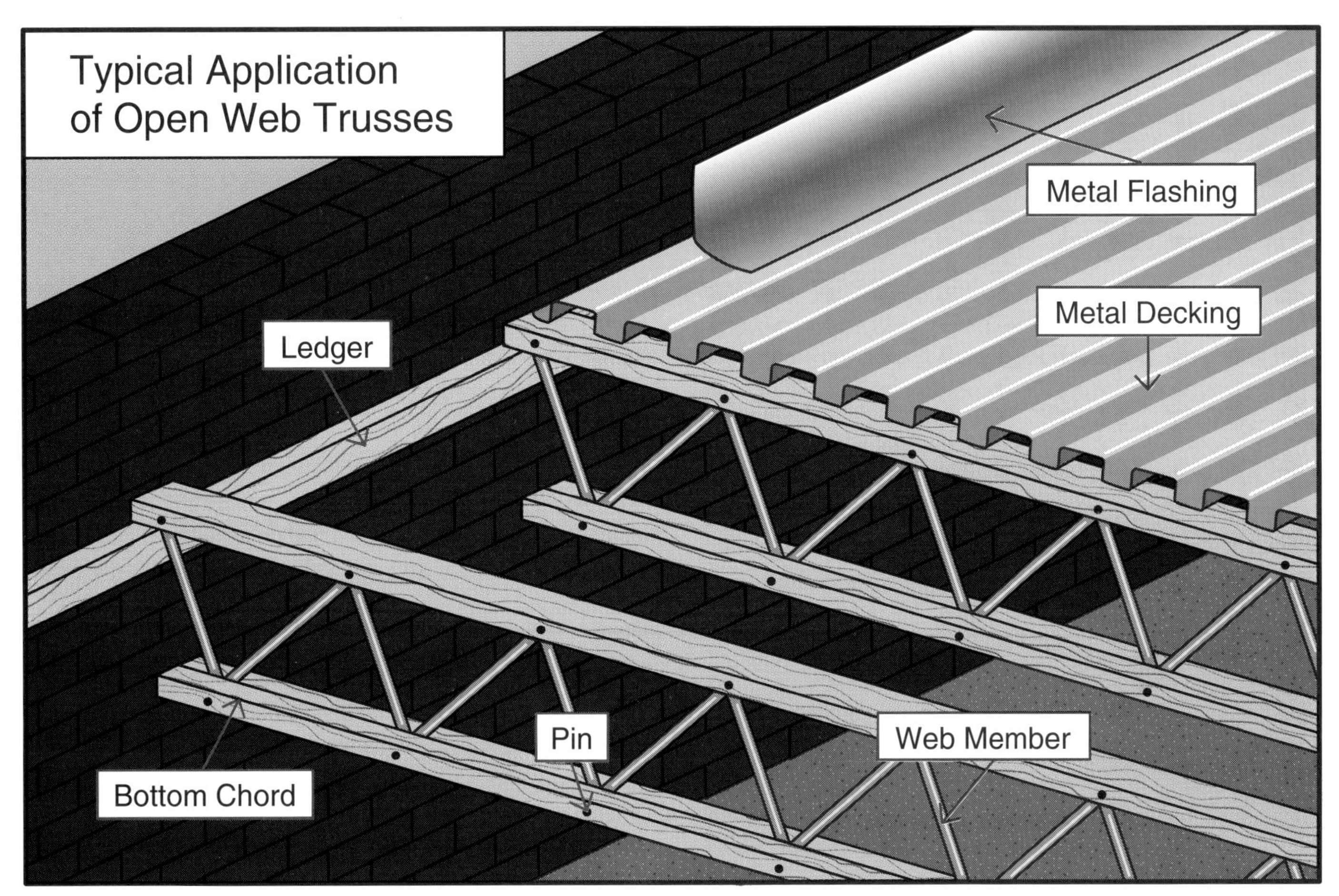

Figure 4.117 The bottom chords of open web trusses are unsupported.

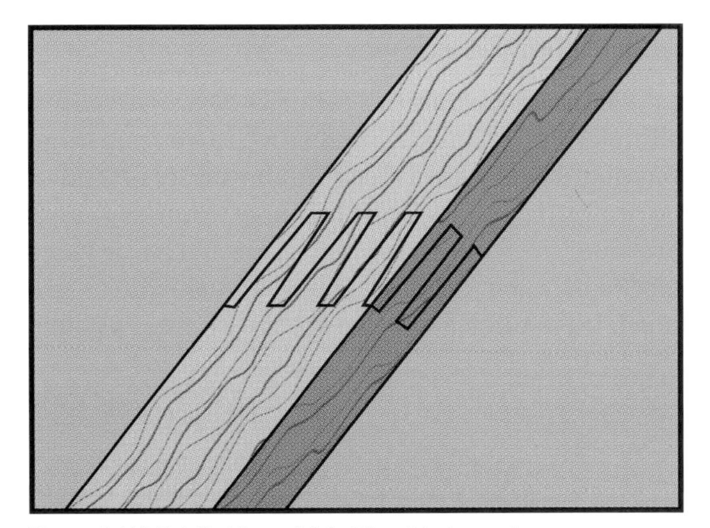

Figure 4.118 Detail of finger-jointed 2- x 4-inch member.

Thus, these trusses offer less fire resistance than conventional roof systems, and early roof collapse is possible. In addition, roof failure can occur when the bottom chord or webbing fails, either from direct fire damage or from connected interior walls falling and pulling them down.

Figure 4.119 Detail of metal gusset plates. Note that these are not corner-nailed.

Figure 4.120 Typical application of metal gusset plate trusses unsupported in the middle of the span. *Courtesy of Wes Kitchel.*

Wooden I Beams

Wooden I beams, used in both roof and floor systems, consist of three main components: the top chord, the bottom chord, and a ⅜-inch plywood or chipboard stem (Figure 4.121). The stem is joined to the top and bottom chords by a continuous, glued joint. The chords are usually made from 2- x 4-inch material, but 2- x 3-inch members are not uncommon. Common spacing for this type of beam is 2 feet (0.6 m) on center, and the area where the roof meets the exterior walls is the strongest point.

Because the stem has relatively little mass, it can burn through and weaken quickly, causing collapse of the beam and the roof or floor system. If heating and air conditioning ducts penetrate the stem, some of the beam's strength may be lost, but more importantly, an avenue for fire spread is created (Figure 4.122).

Figure 4.121 Typical wooden I beams.

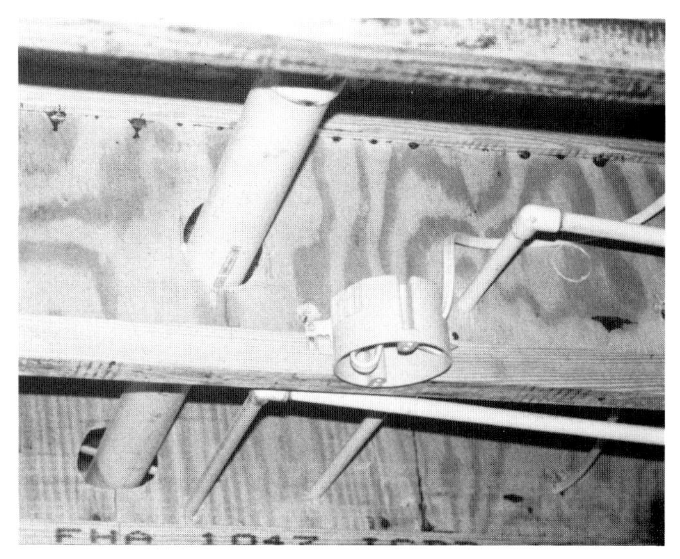

Figure 4.122 Wooden I beams penetrated by plumbing and electrical conduits.

ROOF COVERINGS

Roof coverings are the weather-resistant materials applied over the roof decking or sheathing. They usually consist of one or more layers of underlayment (also called *substrate*) as a vapor barrier and/or insulation over which a weather-resistant covering is laid. The most common substrate is tar paper (also called *roofing felt*). Roof coverings used may be combustible or noncombustible, depending on the application and local code requirements. Roof coverings are classified in NFPA 203, *Guidelines on Roof Coverings and Roof Deck Construction*. The coverings most commonly used are wooden shingles/shakes, composition roofing/shingles, tar and gravel, urethane/isocyanate foam, synthetic membrane, tile/slate, lightweight metal/fiberglass, and steel in a variety of forms.

Wooden Shingles And Shakes

Wooden shingles come in a variety of sizes and are made from a number of different woods. Cedar and redwood are the most common because of their appearance and inherent durability (Figure 4.123). Because these wafer-thin, flat slabs have so little mass, they tend to be highly combustible when dehydrated by age and weather. Consequently, many jurisdictions will allow only fire-retardant-treated shingles to be used.

There are two main differences between wooden shingles and shakes. Shingles are sawn from large rectangular blocks of wood, so the shingles tend to be uniform in shape and thickness. Shakes, which are much thicker than shingles, are split from large blocks of wood, so their shape and thickness are much less uniform (Figure 4.124). Because of their

Figure 4.123 A typical wooden-shingle roof.

Figure 4.124 A typical shake roof showing a less uniform appearance than a shingle roof.

additional mass, shakes are somewhat less susceptible to ignition than shingles, but many jurisdictions still require them to be treated with fire retardant before installation.

Shingles or shakes are usually nailed to wooden 1- x 4-inch or 1- x 6-inch planked sheathing with a space of about 1 inch (25 mm) between the planks. For obvious reasons, this method is called *spaced sheathing* (Figure 4.125). There is usually a single layer of tar paper between the sheathing and the shingles or shakes.

For ventilation purposes, the shingles/shakes can be quickly stripped from the sheathing by inserting the pick of an axe into the space between the planks and pulling the axe laterally in quick, short strokes (Figure 4.126). This procedure alone may provide sufficient ventilation, but if not, the sheathing planks will have to be cut and stripped away also.

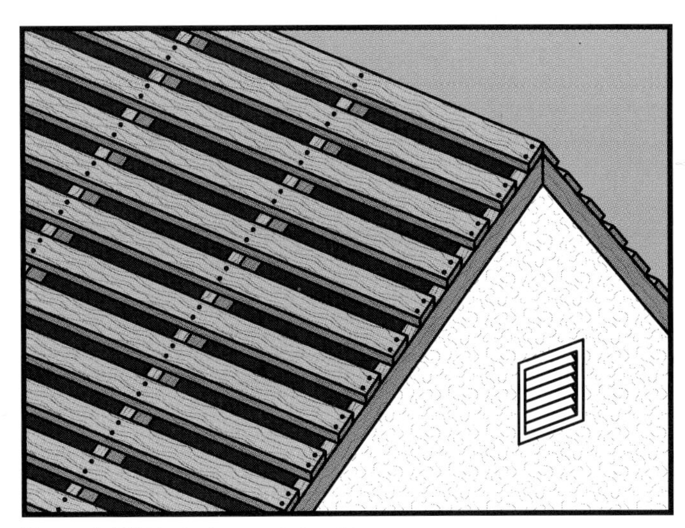

Figure 4.125 Typical spaced sheathing.

Figure 4.126 A firefighter about to strip the roof covering from a roof with spaced sheathing.

Composition Roofing/Shingles

Composition roofing also comes in a variety of materials and shapes. The most common forms are shingles (in rectangular strips) and rolls of various widths, but the 36-inch width is most common (Figure 4.127). They are usually made from a composite (thus, the name) of an asphaltic base material and a granular mineral coating. The mineral coating provides weather resistance and acts as a fire retardant. In recent years, composition roofing containing fiberglass has also become available.

Composition roof coverings are also usually installed over a layer of tar paper. The roof covering is nailed through the tar paper to solid plywood decking or butted plank sheathing (Figure 4.128).

Figure 4.127 Common forms of composition roofing materials.

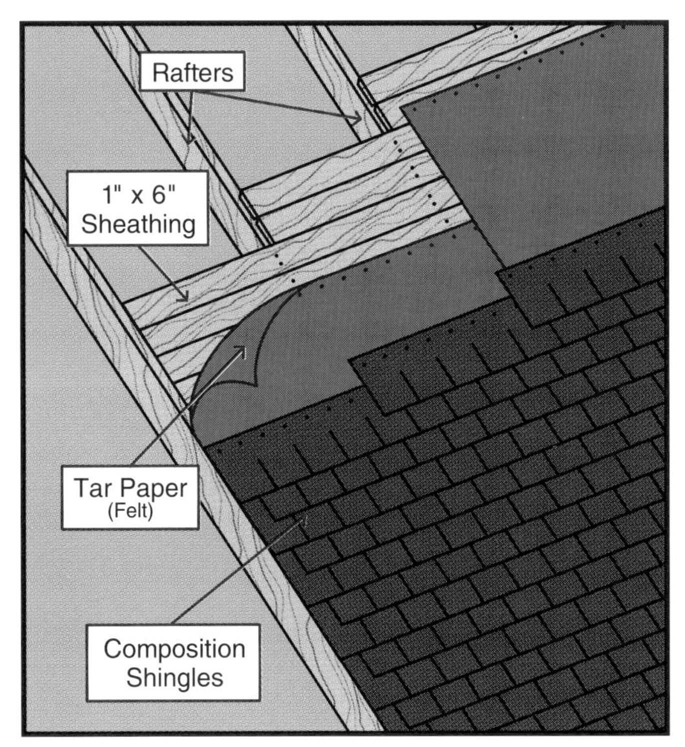

Figure 4.128 Components of a typical composition shingle roof.

While much less combustible than wooden roof coverings, composition roofing materials will burn. When they have burned, they must be stripped from the roof. A flat shovel works well for this operation (Figure 4.129). However, one of the most troublesome aspects of composition roofing for firefighters is that it is a common practice for a new layer of roofing to be applied over an existing one when a building is reroofed. This process can continue over the life of a building, resulting in multiple layers of roofing accumulating on the roof (Figure 4.130). This condition makes the roofing much more difficult to cut for ventilation purposes because this thick asphaltic material tends to gum up the blades of power saws.

Tar And Gravel

Tar and gravel roof coverings (also called *built-up roofs*) are very common on flat or nearly flat roofs. They are found on many types of buildings, ranging from single-family residences to large commercial or industrial buildings. Melted roofing tar is "hot mopped" onto one or more layers of a tar paper substrate over plywood or butted plank sheathing. Pea-sized gravel or crushed slag is most commonly broadcast onto the melted tar to add durability and weather resistance (Figure 4.131),

Figure 4.129 Firefighter using a flat shovel to strip composition shingles from a roof.

Figure 4.130 A typical accumulation of several layers of roofing.

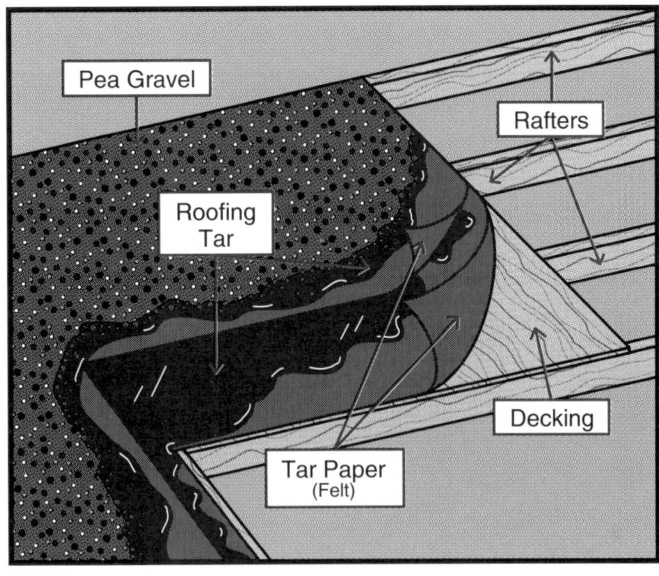

Figure 4.131 Components of a typical tar and gravel or "built-up" roof.

but some roofs may have larger rocks on their surface. In any case, the mineral covering should be scraped away before the roof covering is cut. Because the tar is thermoplastic, it will again soften and liquefy when exposed to the heat of a fire.

A single layer of tar and gravel roofing can be cut easily with a power saw or an axe, but this type of roofing may be found in multiple layers on the roofs of older buildings just as with the composition types of roof coverings. Older tar and gravel roofs may also be covered with a thick coating of foamed roofing (see following section), which often has a silver-colored finish layer. In either case, this type of roof covering can be cut and rolled back as a unit, or it may be louver cut.

Urethane/Isocyanate Foam

Urethane/isocyanate foams are applied on roofs in two forms. One is in the form of 4- x 8-foot sheets of foam insulation (Figure 4.132), and the other is in what are called *foamed-in-place* applications. In either application, the foam is covered by one or more layers of a weather-resistant covering. While these foams may be applied to new roofs, they are often applied over older existing roof coverings. The latter may result in an unusually thick roof covering that slows cutting ventilation openings. These thick, heavily insulated roofs tend to hold heat longer than other roof coverings, which increases the likelihood of flashover or backdraft conditions developing. Because of the toxic products these materials liberate when exposed to fire, firefighters should *always* wear SCBA when ventilating these roofs.

Figure 4.132 Sheets of typical foam insulation.

Single-Ply/Synthetic Membrane

Several liquid elastomers are approved for application on new roofs or over existing built-up roofs after the gravel has been scraped away. These coatings are then sealed with a single layer of any of a variety of approved flexible, water-resistant synthetic membranes (also known as *single-ply roofs*). These membranes are made of neoprene, polyvinyl chloride, chlorinated polyethylene, and bituminous sheets reinforced with polyester or fiberglass (Figure 4.133). The sheets are sealed to the substrates below with an adhesive or by heating the underside of the sheet with an electric heat sealer gun or liquefied petroleum gas (LPG) torch. The seams in the membranes are overlapped and welded together by heating in the same way.

From a ventilation standpoint, these single-ply roof coverings can present some serious problems for firefighters. While the membranes and their substrates are easily cut with common ventilation tools, they are highly combustible and liberate toxic products when they burn.

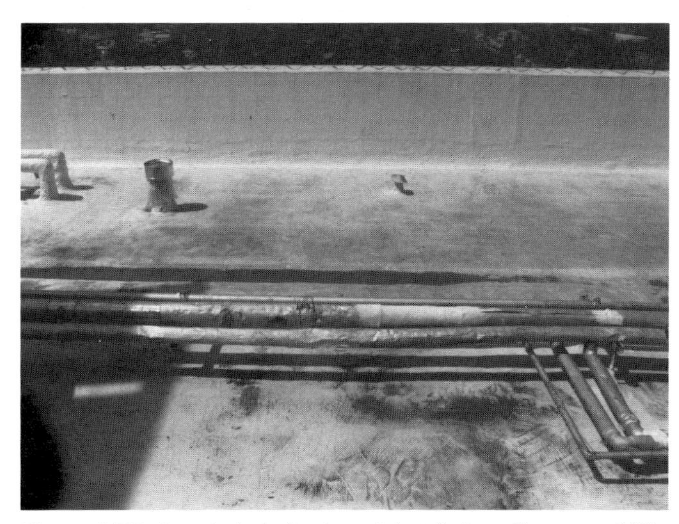

Figure 4.133 A typical single-ply roof installation. *Courtesy of Wes Kitchel.*

Tile/Slate

Spanish tile roofs are common on Spanish-type architecture and are made of curved tiles that are "nested" on the roof, usually over a layer of tar paper and wooden sheathing (Figure 4.134). Concrete or ceramic tile roofs are made of tiles that are flat, interlocking pieces (Figure 4.135). Slate roofs are most common on churches and some larger, usually older residences (Figure 4.136).

Figure 4.134 A typical Spanish tile roof. *Courtesy of Wes Kitchel.*

Figure 4.135 A typical concrete tile roof. *Courtesy of Wes Kitchel.*

Figure 4.136 A typical slate roof common on churches and some larger residences. *Courtesy of Wes Kitchel.*

Tile/slate roofs are extremely fragile and usually cannot be walked on without breaking the tiles, so firefighters walking on the roof may cause considerable damage while getting to the ventilation area. To reduce breakage and to improve footing for the firefighters in these cases, ventilation work should be done from roof ladders.

Ventilating these roofs involves shattering the tiles or slate over the appropriate area. This breakage is best accomplished with a sledgehammer, although a flat-head or pick-head axe can also be used (Figure 4.137). The sheathing can then be cut using an axe or power saw. Broken pieces of tile/slate may slide off the roof creating a safety hazard below, so close coordination between ventilation crews and those working on the ground is critical. In addition, tile/slate roofs carry more weight per unit of surface area than any other roof style and may therefore be susceptible to early roof collapse.

Figure 4.137 A firefighter shattering roof tiles with a sledgehammer.

Lightweight Metal/Fiberglass

This type of roof covering consists of aluminum, fiberglass, or 18- to 20-gauge steel panels over a wooden or metal substructure (Figure 4.138). The panels may be corrugated, ribbed, or shaped to simulate tiles or shakes (Figure 4.139). Buildings with corrugated metal roofs often use plastic or fiberglass panels as skylights in shed and gable roof configurations. The ridge and the area where the roof crosses the outside bearing walls are the strongest points. Because these roof coverings are commonly used over the most lightweight substructures, the roof system will have little fire resistance and is subject to early collapse. Although they can be cut with an axe or a "can opener" (Figure 4.140), using a power saw with a metal-cutting blade is far more efficient.

Steel Clad

In an attempt to secure their property against entry through the roof, some property owners have covered their roofs with steel grids or plates. The entire roof surface, including the steel components, is then resealed with a layer of tar. This procedure sometimes makes the steel difficult to see and the roof nearly impossible to ventilate in a timely manner. Two types of these steel-clad roofs are commonly installed.

Figure 4.138 A typical metal roof.

Figure 4.139 Lightweight steel panels shaped to simulate wooden shakes.

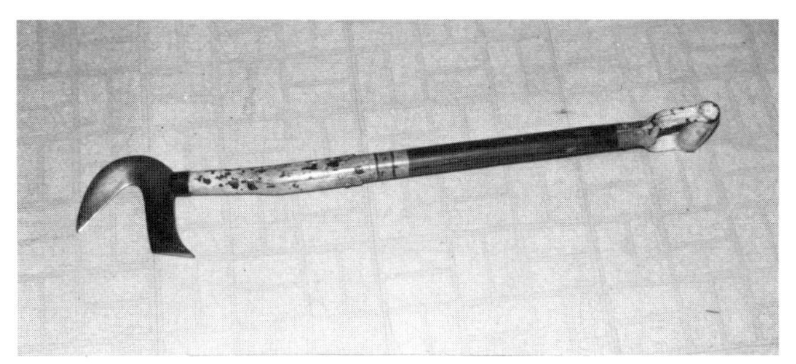

Figure 4.140 A typical "can opener" used to open metal roofs. *Courtesy of Wes Kitchel.*

The first type consists of ¼-inch-thick x 2-inch-wide steel strips laid out on the roof. A second layer of strips crosses the first layer at right angles, and the strips are welded together at each intersection. The grid thus formed leaves small openings of from 6 to 8 inches (150 mm to 200 mm) square. The entire grid is lag-bolted through the roof to the joists below. The entire system is then covered with hot tar to restore weather resistance (Figure 4.141).

Figure 4.141 In a typical installation, the grid of metal strips would be covered with a coating of roofing tar.

The second type consists of laying 4- x 8-foot sheets of steel, ranging from ⅛ to ¼ inch thick, over the entire roof surface. The plates are then welded together to form a continuous layer of steel. Again, hot tar is applied to restore weather resistance.

Steel-clad roofs hold in heat and smoke, promote the development of flashover and backdraft conditions, impede ventilation attempts, and increase the chances of roof collapse due to the added weight. Because ¼-inch steel plate weighs about 10 pounds per square foot (49 kg/m^2), a 20- x 50-foot (7 m by 17 m) roof would have approximately 10,000 pounds (4 536 kg) added to it. The only effective way to prevent this type of construction from interfering with ventilation operations is to learn of its existence through pre-incident inspection and then to develop procedures for dealing with it.

The buildings most likely to have these measures will be high-value occupancies, although in many areas they will also be found on ordinary commercial and industrial buildings. Because it may also discourage potential burglars, some building owners/occupants may be willing to post a prominent sign on the front of the building stating that the building has a steel-clad roof. This sign can alert the incident commander during a fire so that the steel roof protection can be taken into account during size-up and planning.

OPENING THE ROOF

The location for the exit opening will generally be determined by several factors such as a familiarity with the roof structure, the layout of the building, an initial size-up at the scene, and radio communications with firefighters inside. Other aids to locating the spot for the exit opening are smoke coming from existing openings, smoke or fire coming through the roof covering, blisters in the roof covering, or a sagging roof. After determining the best place for the ventilation hole, firefighters can begin opening the roof.

Once the need to open the roof has been determined and the location for the opening has been selected and if there are no existing openings or if they are too small or in the wrong location, firefighters should not hesitate to cut a hole to open the roof (Figure 4.142). However, in most departments, this decision is reserved to the officer in charge of the vent group in coordination with the IC.

Before working on the roof, firefighters should establish primary and alternate escape routes from the roof. Once these routes are established, firefighters should read the roof from a position of safety and sound the roof with a hand tool to make certain that the roof is stable before stepping off the ladder or wall onto the roof. They should continue reading and sounding the roof as they walk on it.

CAUTION: Firefighters should never walk diagonally across a roof but should walk only over the main roof support members.

Key factors in opening roofs safely and efficiently are knowing the location and extent of the fire and the location and direction of the rafters.

Figure 4.142 Firefighters acting on the decision to ventilate.

Most roof assemblies employ a parallel rafter system with rafters spaced from 12 to 24 inches (300 mm to 600 mm) on center, spanning the shortest distance between bearing walls (Figure 4.143). Exceptions to this rule are panelized roofs and specialty roofs using wood or metal trusses or laminated beams. If all other signs indicate that the roof is stable but sounding does not reveal the location and direction of the rafters, firefighters may cut through the roof covering with a chain saw

Figure 4.143 Typical pitched roof framing.

or rotary saw at an angle of 45 degrees to any exterior wall (Figure 4.144). The blade will usually encounter a rafter before the cut is 3 feet (1 m) long. If, after the rafter is located, its direction is still unknown, simply make a small 8- to 10-inch (200 mm to 250 mm) triangular cut at the known rafter location and expose the rafter (Figure 4.145).

Figure 4.144 Making a cut through the roof covering at a 45-degree angle to any exterior wall will help locate a rafter.

Figure 4.145 A small triangular cut will help determine the direction a rafter runs.

When firefighters reach the point where the opening is to be made, every cutting and pulling operation should allow them to continually work back toward an area of safety (Figure 4.146). Once the first cut is made in the roof, the firefighters should avoid putting themselves between that first cut and the seat of the fire. The following precautions should be followed when ventilating any roof:

- Do not make the opening between a crew member and the escape route or in the normal path of travel.
- Cut the roof covering and decking, not the supporting structural members.
- Begin the opening on the leeward side and work against the wind whenever possible so that the smoke and hot gases are blown away from the ventilation crew.
- Remove the ceiling below the roof to ensure adequate ventilation.

Cutting The Hole

As previously mentioned, before the exit opening is cut, the location of the fire should be determined and the ventilation exit opening made as close to the seat of the fire as can be done safely. Ideally, the exit opening should be created directly over the seat of the fire; however, this procedure can be *extremely dangerous*. If there is any doubt about the structural integrity of the roof, only those operations that can be performed from an aerial device should be attempted (Figure 4.147).

When an exit opening is cut, it should be cut large enough to do the job because the time required to enlarge it or to cut several smaller holes is greater than that for making one large one. For example, one 8- x 8-foot (2.4 m by 2.4 m) hole is equal in area to four 4- x 4-foot (1.2 m by 1.2 m) holes (Figure 4.148). When cutting through a roof, crews should make the opening square or rectangular and at right angles to the bearing walls to facilitate repairs and to increase firefighter safety.

Although one large hole is better for ventilation than several small holes, the holes do not necessarily have to be square. Even though firefighters may use a power saw to cut the traditional large, square hole, they still have to contend with the nails in two

Figure 4.146 Firefighters working back toward an area of safety.

Figure 4.147 Firefighters tethered to the aerial device for safety.

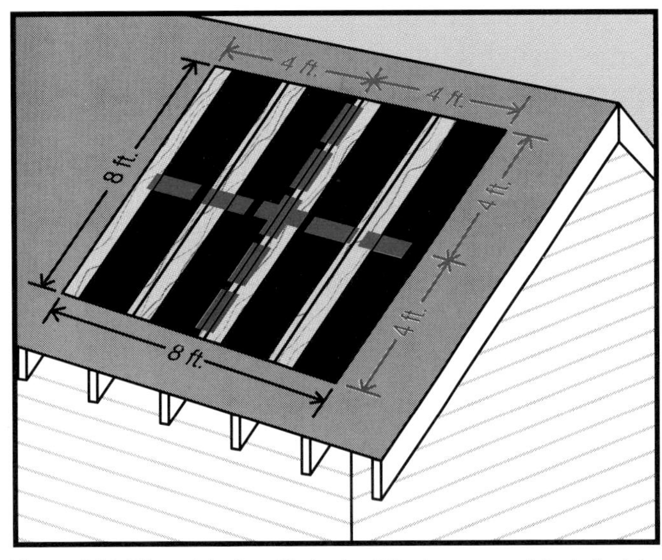

Figure 4.148 The original vent hole should be large enough to do the job.

or more rafters in order to pull the sheathing (Figure 4.149). This task can be a slow and very physically taxing procedure. On the other hand, rectangular, "center-rafter" cuts (louver vents) (Figure 4.150) can be made very quickly and with much less effort. If a larger opening is needed, the hole can be lengthened and used for trench ventilation (described later in this chapter), or a closely grouped series of holes can be cut.

Figure 4.149 Plank sheathing can sometimes be difficult to strip.

Figure 4.150 A typical center-rafter cut or louver vent.

Louver Vents

Cutting louver vents is usually the fastest and most efficient way of opening a roof. Unlike large, square openings, in louver vents there is always a rafter in the center of the sheathing being cut so that a constant reference point is maintained with the rafters. Knowing the location and direction of rafters is critical to safe and efficient roof cutting operations.

Basically, there are two methods that have proven effective for cutting louver vents in roofs. Both methods employ the center-rafter principle and take advantage of the roof's construction to facilitate opening the roof. In the first and most common method, the longest cuts are made parallel to the rafters; in the other method, the longest cuts are made across the rafters.

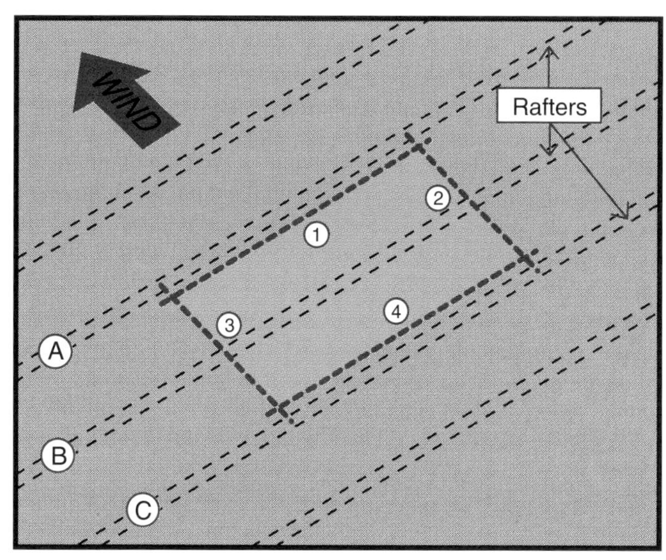

Figure 4.151 The recommended sequence and pattern of cuts for a louver vent.

CUTTING PARALLEL TO THE RAFTERS

The first of these methods involves cutting a long, rectangular center-rafter or louver vent. This method is sometimes referred to as "cutting with the rafters." With a rafter spacing of 2 feet (0.6 m) on center, the hole will be approximately 4 feet (1.2 m) wide and 4 or more feet (1.2 m or more) long (Figure 4.151).

The following steps outline the procedure for cutting parallel to the rafters:

Step 1: Make the first cut parallel to rafter "A," cutting as close to the rafter as possible without cutting it.

Step 2: If the roof is sheathed with plywood or if there are multiple layers of roof covering, make the second cut by starting at rafter "A," crossing over rafter "B" without cutting it, and stopping just short of rafter "C." (NOTE: If the roof covering is thin and if the roof is sheathed with plank sheathing, it may not be necessary to make this end cut because the sheathing provides a break at each joint.)

CAUTION: Care should be taken not to cut into the rafters because this will weaken the roof.

Step 3: Make the third cut by starting at rafter "A," crossing over rafter "B" without cutting it, and stopping at rafter "C." (NOTE: Again, this end cut may not be necessary if you are dealing with a thin roof covering over plank sheathing.)

Step 4: Make the fourth and final cut parallel to rafter "C," cutting as close to the rafter as possible without cutting it.

Step 5: Either push down on the near edge of the louver, pull the far edge to fold back the sheathing, or do both (Figure 4.152).

Step 6: If it is necessary to enlarge the hole, repeat either Steps 1, 2, and 4 or Steps 1, 3, and 4 depending on whether the original hole is enlarged to the right or to the left.

Figure 4.152 Firefighters pushing and pulling a section of sheathing to open a louver vent.

This procedure works equally well on pitched or flat roofs, particularly in newer buildings using 2- x 4-inch trusses that are normally set at 24-inch (600 mm) centers.

CUTTING ACROSS THE RAFTERS

The second method of louver venting, sometimes referred to as "cutting across the rafters," involves making the two longest cuts across the rafter pattern. Subsequent cuts are made parallel to the rafters but in the center between the rafters instead of adjacent to them (Figure 4.153).

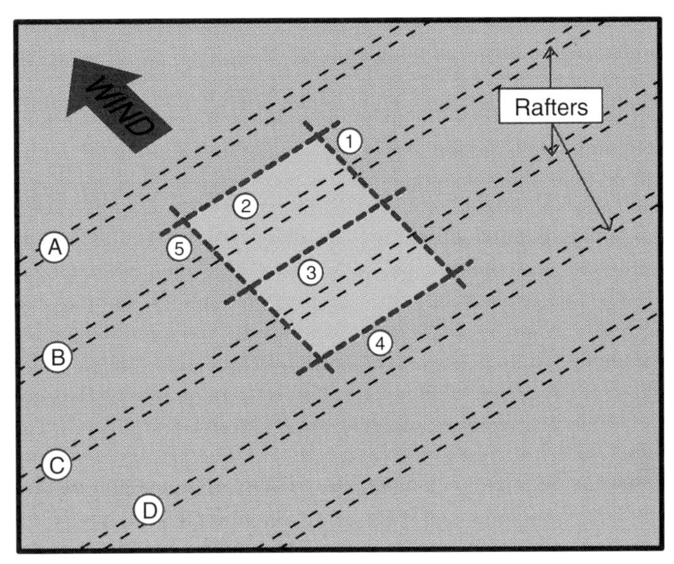

Figure 4.153 The recommended sequence and pattern of cuts for an across-the-rafters cut.

The following steps outline the procedure for the cutting-across-the-rafters method:

Step 1: If the roof is sheathed with plywood or if there are multiple layers of roof covering, start the first cut centered between rafters "A" and "B." Cut across rafters "B" and "C" without cutting them, stopping halfway between rafters "C" and "D." (NOTE: If the covering is thin and the roof is over plank sheathing, you may not need to make this cut because the sheathing provides a break at each joint.)

CAUTION: Care should be taken not to cut into the rafters as this will weaken the roof.

Step 2: Make the second cut about 4 feet (1.2 m) long, centering it between rafters "A" and "B."

Step 3: Make the third cut centered between rafters "B" and "C," making it the same length as the second cut.

Step 4: Make the fourth cut centered between rafters "C" and "D," making it the same length as the second and third cuts.

Step 5: Make the fifth and final cut in the same manner as the first cut, except make it at the opposite ends of the second, third, and fourth cuts.

Step 6: Once all of the cuts are completed, tilt the sheathing between the cuts to open the louvers (Figure 4.154).

NOTE: Regardless of the cutting tool used, it is generally best to complete as many cuts as possible before pulling or folding back the sheathing. This procedure will help reduce the vent group's exposure to heat and smoke during the cutting operations.

Step 7: To enlarge the opening, repeat either Steps 1-4 or Steps 2-5 to widen it, or repeat Steps 1, 2, 3, and 5 to lengthen it (Figure 4.155). In either case, repeat Step 6 to tilt the louvers.

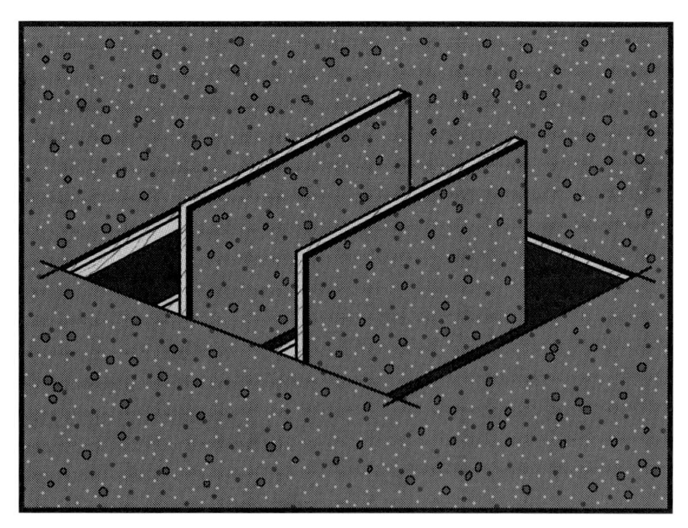

Figure 4.154 Completed louver vents.

Trench (Strip) Ventilation

Trench ventilation, a defensive form of ventilation, is sometimes referred to as *strip ventilation*. In some jurisdictions the term *strip vent* is used to describe a *louver vent*; however, the terms *strip* and

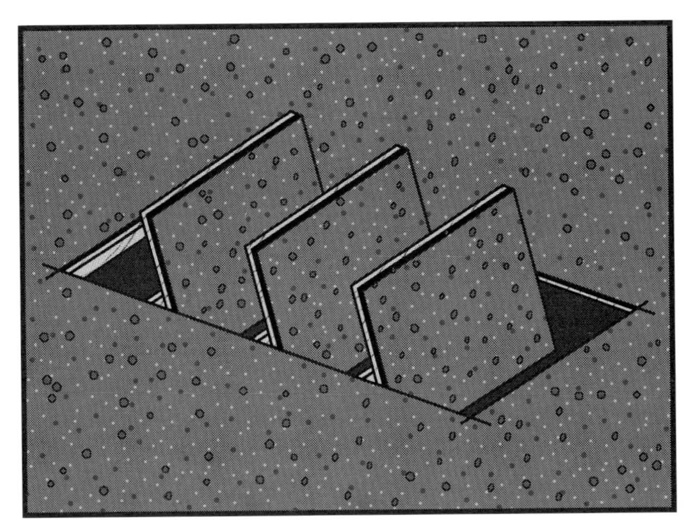

Figure 4.155 An additional vent cut adjacent to the original in order to enlarge it.

trench will be used interchangeably in this manual to indicate a narrow exit opening that extends from one outside wall to the other. The term *trench* may be somewhat misleading. If firefighters infer a long, DEEP hole and cut the rafters, purlins, or other supporting members, the roof system may be seriously weakened in the process. Rather than making a deep cut, firefighters should cut only the roof covering and sheathing and not the members supporting the roof when cutting a hole for trench ventilation.

When done properly, trench ventilation will often help confine a fire to one section of a building by preventing the horizontal spread of heat, smoke, and fire. Trench ventilation has proven to be particularly effective in cutting off running attic fires in buildings with long, narrow, undivided attics.

Trench ventilation is accomplished by making an opening in the roof, approximately 4 feet (1.2 m) wide, across the entire width of the building (Figure 4.156). If the fire is not confined to the attic, the ceiling below must also be breached once the roof has been opened. In most cases, a center-rafter (louver) cut is fastest for making the trench vent.

CAUTION: Because cutting a strip vent takes some time, it is imperative that the trench location be selected far enough ahead of the advancing fire that sufficient time will be available to permit the completion of the opening before the fire reaches the trench.

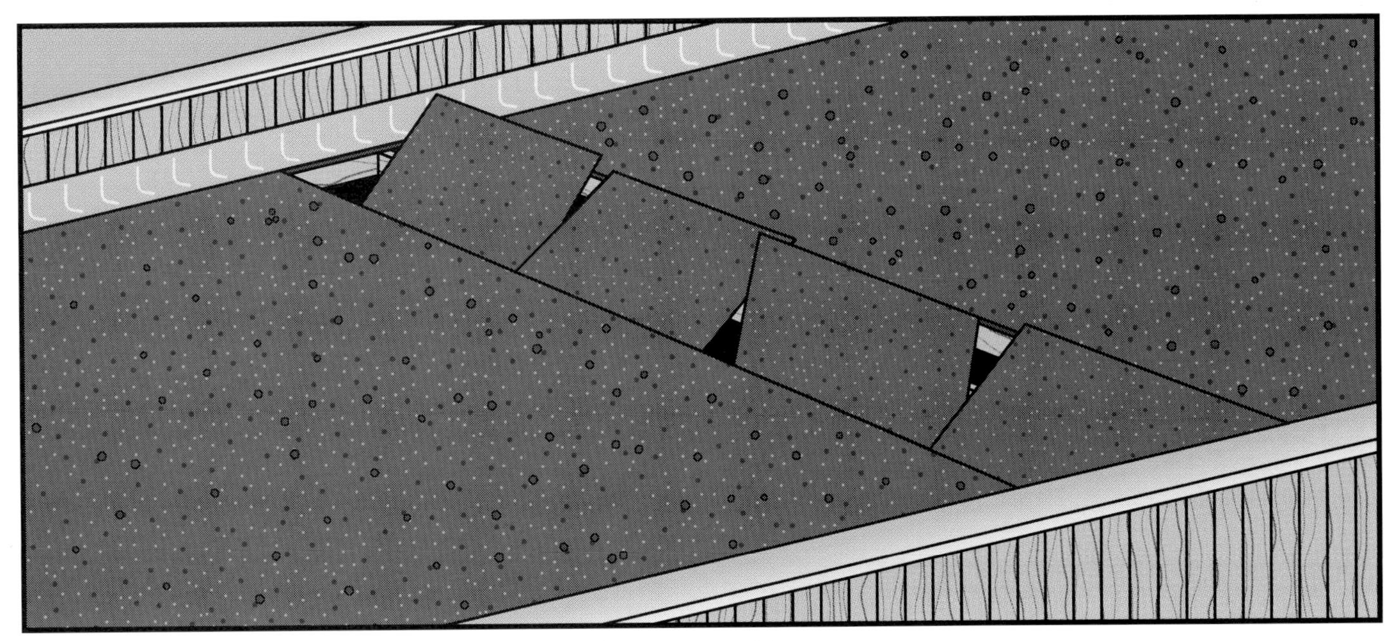

Figure 4.156 A series of louver vents connected to form a trench vent across a flat roof.

Chapter 4 Review

Directions

The following activities are designed to help you comprehend and apply the information in Chapter 4 of **Fire Service Ventilation,** Seventh Edition. To receive the maximum learning experience from these activities, it is recommended that you use the following procedure:

1. Read the chapter, underlining or highlighting important terms, topics, and subject matter. Study the photographs and illustrations, and read the captions under each.
2. Review the list of vocabulary words to ensure that you know the chapter-related meaning of each. If you are unsure of the meaning of a vocabulary word, look the word up in the IFSTA **Orientation and Terminology** glossary or a dictionary, and then study its context in the chapter.
3. On a separate sheet of paper, complete all assigned or selected application and review activities before checking your answers.
4. After you have finished, check your answers against those on the pages referenced in parentheses.
5. Correct any incorrect answers, and review material that was answered incorrectly.

Vocabulary

Be sure that you know the chapter-related meanings of the following words:

- abutment *(116)*
- aesthetic *(98)*
- arduous *(91)*
- ascend *(86)*
- composite *(94)*
- compression *(102)*
- corrugated *(100)*
- deflect *(91)*
- dehydrated *(122)*
- delaminate *(108)*
- egress *(81)*
- elapse *(81)*
- escalation *(118)*
- exothermic *(94)*
- imminent *(88)*
- impede *(108)*
- imperative *(133)*
- impingement *(109)*
- infer *(133)*
- inherent *(122)*
- integrity *(89)*
- intersect *(88)*
- interstitial *(106)*
- interval *(118)*
- inverted *(106)*
- laminated *(98)*
- mercantile *(106)*
- mitered *(120)*
- mitigating *(104)*
- obstruction *(83)*
- pertinent *(89)*
- prefabricated *(119)*
- prone *(102)*
- shard *(104)*
- simulate *(126)*
- stability *(88)*
- susceptible *(108)*
- tension *(102)*
- terminate *(96)*
- versatile *(94)*
- void *(103)*
- wieldy *(108)*

Application Of Knowledge

1. Survey buildings in your jurisdiction and find examples of each of the different pitched, flat, and arched roof types. Note those that would pose potential ventilation problems. *(Local protocol)*
2. List five specific buildings in your jurisdiction constructed with lightweight roofs. *(Local protocol)*
3. Survey buildings in your jurisdiction. Find examples of each of the different roof coverings. Note those that would pose potential ventilation problems. *(Local protocol)*

Review Activities

1. Briefly identify each of the following construction terms:
   - atria *(96)*
   - bearing wall *(98)*
   - bulb tee *(110)*
   - buttress (noun) *(116)*
   - chord *(116)*
   - collar beam *(100)*
   - crawl space (cockloft, interstitial space) *(106)*
   - flashing (noun) *(96)*
   - girder *(115)*
   - gusset plate *(104)*
   - hatch (noun) *(96)*
   - I beam *(110)*
   - isocyanate foam *(125)*
   - joist *(98)*
   - laminated beam *(98)*
   - liquid elastomer *(125)*
   - louver *(94)*
   - parapet *(85)*
   - partition *(120)*
   - pilaster *(118)*
   - purlin *(118)*
   - rafter *(87)*
   - ridge *(98)*
   - roof pitch *(98)*
   - saddle *(118)*
   - thermoplastic *(96)*
   - tie rod *(102)*
   - truss *(98)*
   - turnbuckle *(116)*
   - underlayment (substrate) *(122)*
   - urethane foam *(125)*
   - roof valley *(101)*
   - web member *(102)*
   - wooden shake *(100)*
2. Define *vertical ventilation. (83)*
3. State the purpose of vertical ventilation. *(83)*
4. List hazards associated with vertical ventilation. *(83, 84)*
5. Explain "dead load" versus "live load" in the context of vertical ventilation. *(85 & 86)*

6. List safety precautions that firefighters should observe when vertically ventilating a building. *(84, 85)*

7. Explain what is meant by the expression "reading the roof." *(85 & 87)*

8. List factors that a ventilation officer must consider before, during, and after vertically ventilating a building. *(87)*

9. Explain how and why a firefighter "sounds" a roof. *(87, 88)*

10. Draw simple lines and arrows to illustrate the following orientation terms:
    - diameter *(94)*
    - 45° angle *(98)*
    - horizontal *(83)*
    - lateral *(105)*
    - parallel *(104)*
    - perimeter *(92)*
    - perpendicular *(88)*
    - vertical *(83)*

11. Explain when it is necessary to have a charged hoseline on the roof during vertical ventilation operations. *(89)*

12. State vertical ventilation uses for each of the following tools:
    - pick-head axe *(91 & 94)*
    - rotary saw *(91)*
    - chain saw *(92)*
    - burning bar *(94)*
    - pike pole *(94, 95)*
    - rubbish hook *(95)*
    - sledgehammer *(95)*

13. Describe each of the following existing roof openings:
    - scuttle hatch *(96)*
    - bulkhead *(96)*
    - skylight *(96)*
    - monitor *(97)*
    - light and ventilation shafts *(97, 98)*

14. Create a simple drawing to illustrate each of the following pitched roof types:
    - bridge truss *(102)*
    - butterfly *(103)*
    - gable *(100)*
    - gambrel *(103)*
    - hip *(100)*
    - lantern *(101)*
    - mansard *(102)*
    - modern mansard *(103 & 110)*
    - sawtooth *(103)*
    - shed *(102)*

15. Explain how specific construction features of each of the pitched roof types affect vertical ventilation operations. *(101-104)*

16. List hazards associated with vertically ventilating pitched roofs. *(104)*

17. Describe general procedures for vertically ventilating pitched roofs. *(104, 105)*

18. Briefly identify each of the following types of flat roofs:
    - concrete *(110)*
    - inverted *(106)*
    - metal deck *(108)*
    - panelized *(112)*
    - poured gypsum *(110)*
    - rain roof *(107)*
    - wooden deck *(108)*

19. Explain how specific construction features of each of the flat roof types affect vertical ventilation operations. *(106-110)*

20. List hazards associated with vertically ventilating flat roofs. *(112)*

21. Describe general procedures for vertically ventilating flat roofs. *(113-115)*

22. Briefly identify each of the following types of arched roofs:
    - bowstring *(116)*
    - lamella *(117)*
    - ribbed (trussed) *(116)*

23. Explain how specific construction features of each of the arched roof types affect vertical ventilation operations. *(116, 117)*

24. List hazards associated with vertically ventilating arched roofs. *(117)*

25. Describe general procedures for vertically ventilating arched roofs. *(118)*

26. Define *lightweight construction. (118)*

27. Identify construction characteristics of each of the following types of lightweight roofs:
    - panelized *(118)*
    - open web trussed *(119)*
    - metal gusset plate trussed *(120)*
    - wooden I beam *(122)*

28. List ventilation techniques and hazards associated with each type of lightweight roof construction listed in Activity 27. *(118-122)*

29. Describe each of the following types of roof coverings:
    - wooden shingles *(122, 123)*
    - wooden shakes *(122, 123)*
    - composition roofing/shingles *(123, 124)*
    - tar and gravel *(124, 125)*
    - urethane/isocyanate foam *(125)*
    - single ply, synthetic membrane (elastomer) *(125)*
    - tile/slate *(125, 126)*
    - lightweight metal/fiberglass *(126, 127)*
    - steel clad *(127)*

30. Describe ventilation techniques and hazards associated with each type of roof covering listed in Activity 29. *(122-128)*

31. List general precautions for performing vertical ventilation on any roof. *(128, 129)*

32. Briefly describe each of the following types of ventilation openings:
    - regular square/rectangular *(129, 130)*
    - louver ("center-rafter") *(131)*
    - trench (strip) *(132, 133)*

5 Forced Ventilation

This chapter provides information that addresses the following performance objectives of NFPA 1001, *Standard for Fire Fighter Professional Qualifications* (1992):

Chapter 3 — Fire Fighter I

3-9.3 Describe the advantages and disadvantages of the following types of ventilation:
- (a) Vertical
- (b) Horizontal
- (c) Trench/strip
- (d) Mechanical
- (e) Mechanical pressurization
- (f) Hydraulic

3-9.10 Define procedures for the types of ventilation referred to in section 3-9.3.

Chapter 4 — Fire Fighter II

4-9.1 Identify the manual and automatic venting devices found within structures.

4-9.4 Identify the location of the opening, the method to be used, and the precautions to be taken when ventilating a basement.

4-9.5 Identify fire ground situations where forced ventilation procedures may be required.

Safety Points

In its discussion of forced ventilation, this chapter addresses the following safety points:

- Incident commanders must be aware of the advantages and disadvantages of using forced ventilation as well as the risks involved.
- Firefighters should know the capabilities and limitations of forced ventilation in an involved structure.
- Incident commanders and firefighters must be aware of the effects of misapplying forced ventilation.
- Firefighters must be aware of the dangers presented by rapidly-turning fan blades in operating blowers and smoke ejectors.
- Firefighters must be wary of walking into smoke ejectors placed in doorways or in smoke-filled rooms.
- To reduce the chances of a smoke ejector becoming a source of ignition in a contaminated atmosphere, the firefighter should turn the switch on the ejector on *before* plugging the unit into the power source.
- Firefighters must be aware of the potential for oxygen deficiency and/or the presence of heavier-than-air gases in basements and other below-grade spaces.
- Firefighters must use SCBA when in a contaminated or oxygen-deficient atmosphere.
- Firefighters should maintain a familiarity with the occupancy, contents, and equipment in buildings within their response district.
- Firefighters should not attempt to operate built-in air-handling equipment in an involved structure but should rely on the building engineer to do so under their direction.
- Firefighters must use only smoke ejectors with explosion-proof motors whenever ejectors must be operated in potentially flammable atmospheres.
- To avoid spreading the fire to uninvolved areas, and perhaps jeopardizing occupants or other firefighters, firefighters should know the location and extent of the fire before starting positive-pressure ventilation (PPV).
- Any form of ventilation, but especially PPV, must be executed as part of a coordinated effort working with the other parts of the operation.
- Firefighters working within an involved building should not be allowed to remove their SCBA until the interior atmosphere has been sampled and determined to be within safe limits.

Chapter 5
Forced Ventilation

INTRODUCTION

Ventilation efforts should be in concert with existing atmospheric conditions, taking advantage of natural ventilation whenever possible. However, in some situations, natural ventilation may be inadequate and may have to be supplemented or replaced with forced ventilation to provide a tenable atmosphere and to facilitate rescue and suppression operations. Forced ventilation involves the use of fans, blowers, nozzles, and other mechanical devices to create or redirect the flow of air within an involved space to facilitate operations within the space.

Some terms are commonly used (and misused) by many firefighters when discussing forced ventilation theory and methods. For instance, the term ***positive pressure*** is redundant in that pressure is *always* positive; the term ***negative pressure*** is a misnomer because an absence of pressure is a *vacuum*; using the term ***venturi effect*** is not technically accurate to describe the *entrainment* of air being ejected from a confined space. However, because these terms are in such widespread use in the fire service, they will also be used that way in this manual. For purposes of clarity, ***positive pressure*** will refer to *blowing* air from outside a confined space *into* the space, ***negative pressure*** will refer to *ejecting* (exhausting) air *from within* a confined space to the outside, and ***venturi effect*** will also be used to describe air being drawn into the stream of air being mechanically exhausted from a space. In this manual, "outside" may include an area within a building but outside of the space to be ventilated, as well as outside of the building.

This chapter discusses the advantages and disadvantages of forced ventilation, the devices necessary to create forced ventilation, situations in which forced ventilation is needed, and the techniques used in applying forced ventilation. Included in the discussion of forced ventilation devices are heating, ventilating, and air-conditioning (HVAC) systems and other built-in air-moving equipment.

ADVANTAGES AND DISADVANTAGES OF FORCED VENTILATION

There are some distinct advantages of using forced ventilation in terms of both control and effectiveness. However, there are also some clear disadvantages in the resources required for its use and in its application.

Advantages

Using forced ventilation eliminates or reduces the effect of unstable and erratic winds on ventilation efforts within a structure. By providing a dependable, controllable airflow, greater control of the fire situation is possible. This airflow allows the flames, heat, smoke, and gases to be directed away from building occupants and away from uninvolved portions of the structure (Figure 5.1). Forced ventilation can channel the airborne products of combustion or other contaminants out of a building by the most efficient and least destructive path and allow fresh air to be reintroduced into the space. In both fire and nonfire situations, using proper forced ventilation in conjunction with natural ventilation allows a tenable atmosphere to be restored faster and more efficiently than with natural ventilation alone.

Some of the advantages of using properly applied forced ventilation in fire and nonfire situations are that it accomplishes the following:

- Creates a safer interior operating environment
- Aids in search and rescue
- Aids in locating the source of the problem
- Speeds removal of airborne contaminants
- Improves visibility
- Reduces interior heat levels
- Speeds restoration of a tenable atmosphere
- May supplement natural airflow
- Reduces smoke and fire damage
- Promotes good public relations

In fire situations, reducing heat and smoke within the structure increases safety for firefighters and improves survivability for any occupants still inside. As fresh air is drawn in to replace the smoke being ejected, temperatures within the space drop, and vision is enhanced because the smoke and steam are channeled out (Figure 5.2). With properly applied forced ventilation techniques, these changes can be accomplished very quickly. The

Figure 5.1 Heat, smoke, and gases being forced out of a structure.

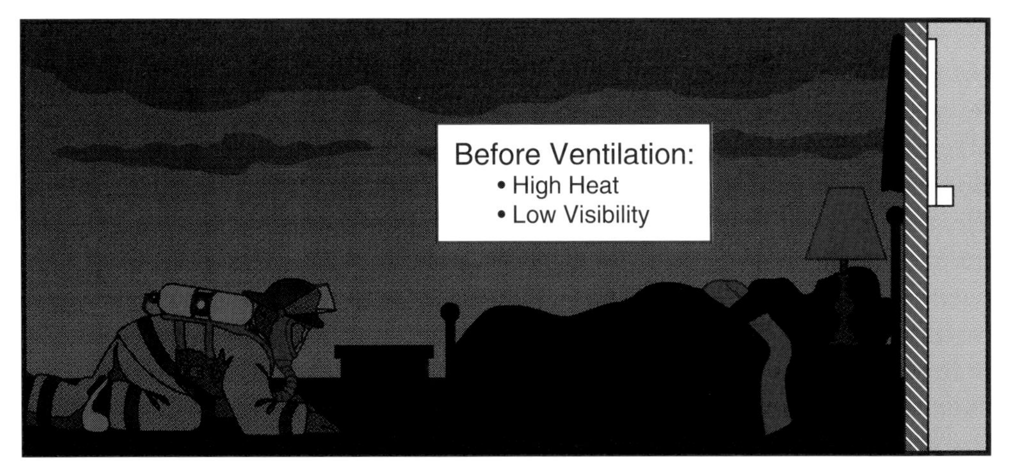

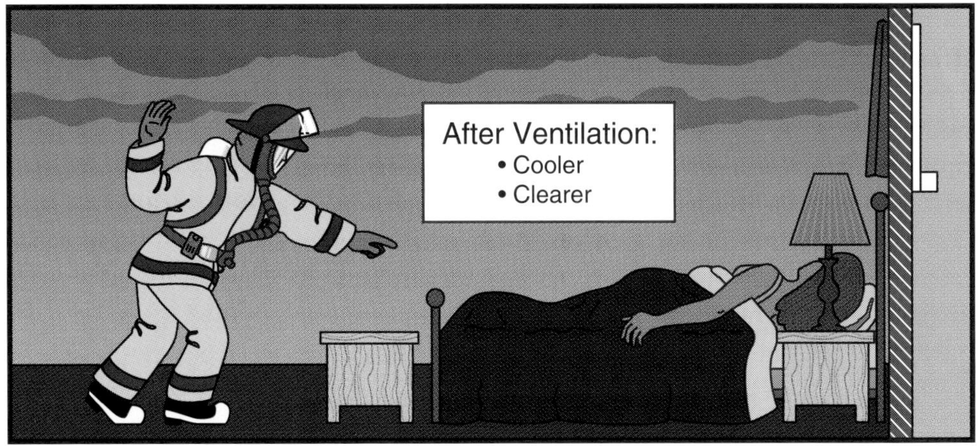

Figure 5.2 The before-and-after effects of ventilation.

reduction in interior temperature is especially important to firefighter safety because the protective clothing worn by structural firefighters traps so much body heat that fatigue and heat exhaustion can develop quickly if the ambient temperature is high (Figure 5.3). Firefighter safety is also greatly enhanced when smoke and steam are ejected and not allowed to obscure vision by filling the space from ceiling to floor.

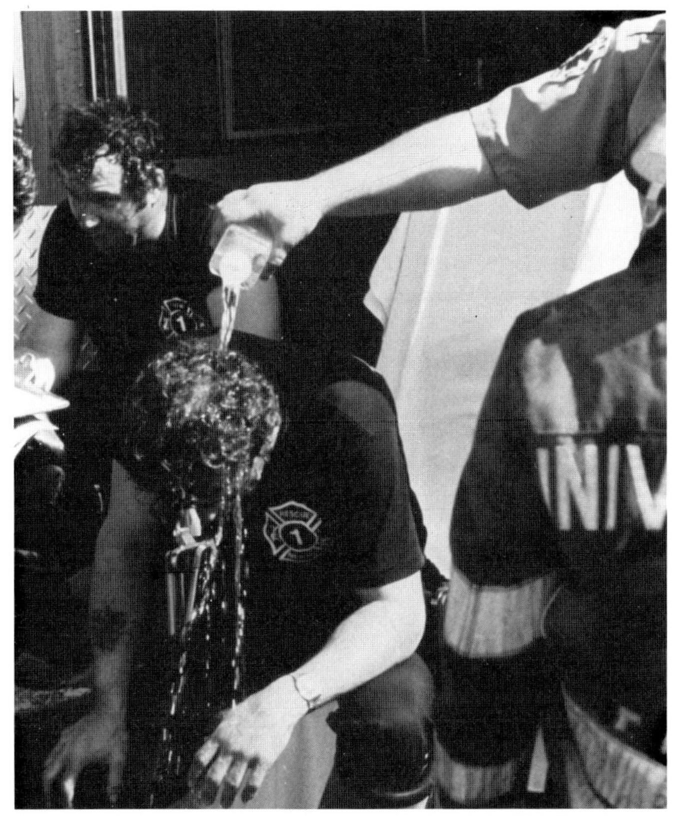

Figure 5.3 A firefighter overcome by heat and exhaustion. *Courtesy of Ron Jeffers.*

In situations where the wind is not blowing — or is minimal — forced ventilation can mechanically create all the airflow necessary to adequately ventilate the space. It will also allow the atmosphere within the space to be cleared faster than with natural ventilation alone. When contaminants are exhausted to the outside, the atmosphere inside will be made tenable at the same rate by the fresh replacement air; thus, the faster the contaminants are ejected, the faster a tenable atmosphere will be restored.

When heat and smoke are ejected from a structure at or near the seat of the fire, flame spread and smoke damage are confined to the smallest possible area (Figure 5.4). Often, this confinement means that the fire will be limited to the amount of involvement that was found when the first fire department units arrived and that smoke damage will be minimized. When the fire is kept relatively small, the amount of water needed for fire extinguishment is also reduced, thus helping to limit water damage.

Anytime that rescues can be facilitated, extinguishment accomplished quickly, fire and smoke damage limited, and water damage kept to a minimum, the overall fire loss will be kept as low as possible. Whenever the fire loss can be kept low, the expense and inconvenience suffered by the property owner will be kept to a minimum. This usually translates into positive public relations for the fire department (Figure 5.5).

Figure 5.4 Smoke and flame being forced out near the seat of the fire.

Figure 5.5 A grateful homeowner shows his appreciation for the firefighters' efforts.

Disadvantages

If forced ventilation is misapplied or inadequately controlled, it can compound the situation it was intended to correct. Improperly controlled forced ventilation can greatly increase the intensity of a fire or spread fire and airborne contaminants throughout a building, endangering occupants and/or firefighters. Improperly applied forced ventilation can reduce the safety of interior operations by obscuring vision, by failing to reduce interior temperatures, and by channeling contaminants into exit passageways.

When compared to natural ventilation, there are other possible disadvantages of forced ventilation. In addition to the personnel needed to create the necessary ventilation openings, blowers or smoke ejectors are needed along with additional personnel to transport, set up, and operate this equipment (Figure 5.6). Smoke ejectors or blowers must be positioned, and any auxiliary equipment, such as hangers or stands, must be assembled. Ideally, these machines should be monitored during operation, requiring even more personnel.

Figure 5.6 Firefighters setting up a smoke ejector.

For this equipment to operate, some source of motive power must be supplied and maintained. Whether the power is electrical, hydraulic, or from a gasoline-driven engine, the continued operation of the equipment is dependent upon that power source being maintained. Electric power must come either from the structure, which may or may not be a safe and dependable source during a working fire, or from generators on the fire apparatus, which may be needed to supply power to lights and/or power tools (Figure 5.7). Using water pressure requires that another line be stretched from the pumping engine to the fan, and this power source is subject to interruption if the apparatus runs out of water or breaks down. Gasoline-driven blowers must eventually be refueled, and safety rules dictate that they must be shut down during refueling (Figure 5.8).

To be used effectively, forced ventilation requires that personnel receive additional specialized training in forced ventilation theory and in the operation of the equipment (Figure 5.9). Personnel must invest the time and effort needed to learn and practice the techniques under controlled conditions before attempting to apply them in the field. If this training is not carried out, or if it is done

Figure 5.7 Smoke ejector cords should always be plugged into the power source outside the contaminated atmosphere.

Figure 5.8 The ventilation process can be interrupted when gasoline-driven blowers are shut down to be refueled.

Figure 5.9 A training officer reviews forced ventilation theory prior to starting an exercise.

inadequately, forced ventilation can be misapplied with disastrous results.

If forced ventilation is misapplied, fires can be intensified, resulting in greater fire loss and making the fire fighting operation more dangerous. If forced ventilation is introduced into a structure before attack lines are ready to be taken in, the fire will be supplied with additional air that will intensify its rate of combustion and increase fire spread (Figure 5.10). If forced ventilation is applied at the wrong location, it can cause the fire to spread to uninvolved portions of the structure, impede suppression efforts by pushing heat and smoke toward firefighters attempting to advance hoselines, and perhaps cut off means of escape for occupants.

Even if forced ventilation is not misapplied, the equipment needed to create it may also represent safety hazards. Firefighters may walk into smoke ejectors hung in doorways or trip over blowers positioned just outside exterior doorways. In most cases, using handline nozzles to eject smoke forces firefighters to remain in the most heavily contaminated area during the ventilation operation.

Figure 5.10 Forced ventilation should not begin before attack lines are in position and ready.

The disadvantages of forced ventilation are summarized in the following list:

- Requires additional personnel to set up and operate
- Is dependent on a power source
- Requires additional training
- May increase property damage (if misapplied)
- May reduce personnel safety (if misapplied)
- May produce negative public relations (if misapplied)

FORCED VENTILATION DEVICES

The most common forced ventilation devices are portable air-moving devices such as smoke ejectors, blowers, and hoseline nozzles. Also included in the category of forced ventilation devices are various forms of fixed air-handling devices such as residential and commercial climate control systems.

Portable Air-Moving Devices

Portable air-moving devices commonly used by fire departments consist of smoke ejectors and the auxiliary equipment necessary to maximize their effectiveness, blowers, and hoseline nozzles.

SMOKE EJECTORS

Smoke ejectors are ducted fans of various sizes, usually in boxlike housings (Figure 5.11). They are most often used inside a structure or other confined space to exhaust smoke or other contaminants to the outside. Smoke ejector sizes are based on the diameter of the fan, usually ranging from 16 to 24 inches (400 mm to 600 mm), and their air-moving capacity is rated in cubic feet per minute (ft^3/min [m^3/min]). Some are driven by gasoline engines, but most are driven by electric motors. Some electrically-driven smoke ejectors are intended to be used in potentially explosive atmospheres and have explosion-proof motors (Figure 5.12). To prevent injury to personnel working near these devices, and to prevent foreign objects from accidentally entering the duct and damaging or being damaged by the ejector fan, the intake and exhaust openings are protected by sturdy wire grilles. Smoke ejectors are sometimes fitted with flexible ducting in order to access certain difficult-to-reach spaces (Figure 5.13).

BLOWERS

Blowers are fans that may or may not be ducted and that are almost always used to blow fresh air from the outside into a confined space to pressurize it and thereby force contaminants out. Blowers tend to be larger in size and rated capacity than smoke ejectors, with 18- to 36-inch (450 mm to 900 mm) sizes being the most common (Figure 5.14). Their rated capacities range from less than 5,000 ft^3/min (142 m^3/min) to more than 16,000 ft^3/min (453 m^3/min). While some electric or water-

Figure 5.14 A typical blower used for ventilation.

Figure 5.11 A typical smoke ejector.

Figure 5.12 A smoke ejector with an explosion-proof motor.

Figure 5.13 A smoke ejector fitted with flexible ducting. *Courtesy of Lisle-Woodridge Fire District.*

powered blowers are available, the majority are powered by gasoline-driven engines. Some large, vehicle-mounted blowers have fans of 48 to 72 inches (1 200 mm to 1 800 mm) and are capable of producing from 80,000 ft^3/min to more than 200,000 ft^3/min (2 265 m^3/min to 5 663 m^3/min) (Figure 5.15). Some blowers come equipped with spray nozzles for injecting water into the airstream, and others may be so adapted. Injecting water into the airstream is an attempt to combine fire suppression with the ventilation effort; however, this technique appears to be of little if any value in fire extinguishment.

Figure 5.15 A vehicle-mounted blower. *Courtesy of Joel Woods.*

HOSELINE NOZZLES

Ordinary adjustable fog nozzles may also be used to direct and control airflow for forced ventilation. While this technique is used most often with handline nozzles (Figure 5.16), it can also be applied by using adjustable master stream nozzles (Figure 5.17). Using the same airflow principle as a properly placed smoke ejector, nozzles can be quite effective when used for hydraulic ventilation.

Figure 5.16 A handline nozzle being used for hydraulic ventilation.

Fixed Air-Handling Devices

Normally used for routine, day-to-day climate control, this type of equipment is installed in a variety of buildings, ranging from residential dwell-

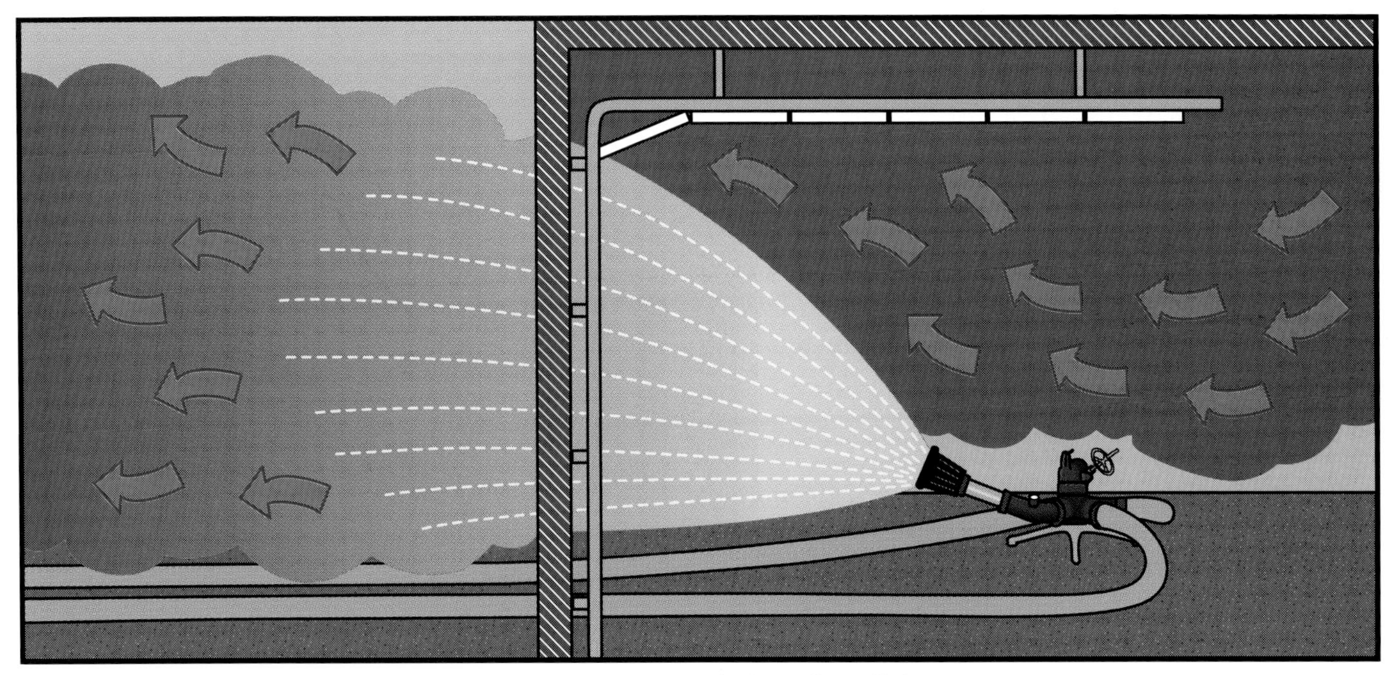

Figure 5.17 A master stream nozzle being directed through a large opening for hydraulic ventilation.

ings to large mercantile or industrial complexes and high-rise apartment and office buildings. The equipment in these buildings is as complex and varied as the occupancies themselves, so the only way that firefighters can be sure of being able to use this equipment to its full advantage is to consult qualified building-maintenance personnel to learn the capabilities and limitations of all such equipment installed in their response district (Figure 5.18). As the occupancy changes, sometimes the building and its equipment change also. Firefighters must revisit these buildings periodically to stay current on any changes that may affect fire behavior or ventilation considerations.

Figure 5.18 A building engineer explains the operation of an HVAC system to firefighters.

RESIDENTIAL CLIMATE CONTROL SYSTEMS

In most cases, these climate control systems serve only to create a need for forced ventilation by circulating smoke throughout the structure, and their design does not allow them to be used for clearing the atmosphere. In a few situations, however, even this equipment can be of some value. For example, if a residential system is designed to draw replacement air from the outside, it can aid in diluting and displacing cold smoke.

COMMERCIAL CLIMATE CONTROL SYSTEMS

In some cases, these systems may be little more than larger versions of the residential equipment (Figure 5.19). In other cases, the systems may be equipped with some basic fire-control features such as fire dampers in the ducts; these dampers are activated by fusible links. This type of system, common to mercantile and light industrial occupancies, usually offers little more assistance to ventilation operations than do the residential systems.

Heavy-duty HVAC systems are found in many major mercantile occupancies, enclosed shopping malls, high-rise apartment and office buildings, and certain heavy industrial occupancies (Figure 5.20). They are often designed with a variety of features that will assist in limiting the spread of fire and in performing forced ventilation. These systems include heat- and/or smoke-activated dampers in the ducting, heat- and smoke-detection systems integrated into the building's alarm system (Figure 5.21), and automatic sprinkler protection. Perhaps the most important feature of HVAC systems for forced ventilation is that the airflow in some of these systems may be manually controlled to assist in exhausting cold smoke (Figure 5.22). For more information on these systems see NFPA 92A, *Recommended Practice for Smoke Control Systems*, and NFPA 92B, *Guide for Smoke Management Systems in Malls, Atria, and Large Areas.*

Figure 5.19 Typical commercial climate control ducting.

Figure 5.20 Typical heavy-duty HVAC machinery.

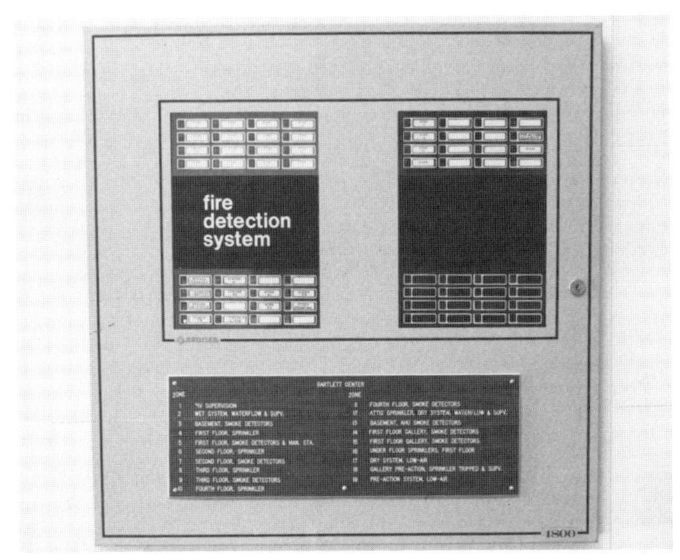

Figure 5.21 HVAC systems may have smoke-detection systems integrated into the building's alarm system.

Figure 5.22 A firefighter working with the building engineer to use the HVAC system to exhaust cold smoke from the building.

TYPES OF FORCED VENTILATION

Forced ventilation can be indicated in a variety of different situations, using a variety of different devices, and depending on the nature of the particular situation, may require different methods of application. Some methods lend themselves to certain situations better than others. In addition, the local department's available resources, level of training, and operational philosophy will combine to influence which ventilation method is used. The following are the most common methods used by fire departments in the United States and Canada.

Positive-Pressure Ventilation (PPV)

A relatively recent development, positive-pressure ventilation (PPV) has evolved into one of the most effective methods of forced ventilation available to firefighters. It involves introducing fresh air into a confined space at a rate faster than it is exiting, thus creating a slight positive pressure within the space (Figure 5.23). This positive pressure counteracts the pressure being generated by the combustion process and/or by adverse winds (Figure 5.24). If PPV is set up properly, it will help confine the fire and prevent the products of combustion from spreading to uninvolved portions of the building.

When compared to the negative-pressure ventilation common in the fire service for many years, PPV is far more effective at moving air within and out of a building and is a much safer operation. PPV is safer for those assigned to set it up because in most cases it can be done from the outside. It is safer for the attack crews because it keeps interior temperatures lower by introducing cool, fresh air, and it improves visibility by limiting the spread of smoke and steam within the building.

Because of its effect on fire behavior, PPV need not be limited to the overhaul phase but can be an integral and extremely effective part of the initial

Figure 5.23 Firefighters using a blower to pressurize a building.

fire attack. If PPV is set up as attack lines are being pulled and positioned for entry into a burning building, the ventilation process can be started as soon as the attack lines are charged and ready (Figure 5.25). While PPV is very effective in restoring a tenable atmosphere following an interior fire, waiting until after knockdown to begin PPV will subject attack crews to unnecessary punishment and perhaps will allow the fire to spread within the building, thereby increasing both smoke and fire damage.

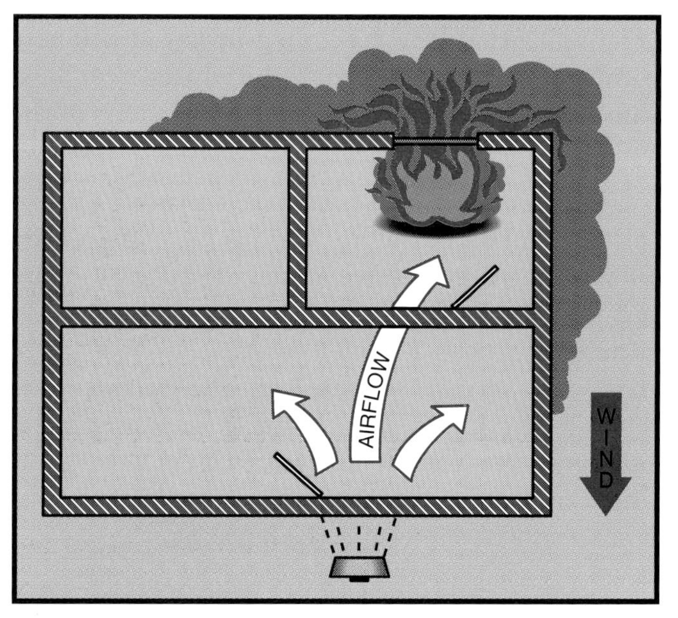

Figure 5.24 PPV overcoming the pressure from adverse winds.

Figure 5.25 An attack team ready to enter the building as soon as PPV is started.

Negative-Pressure Ventilation (NPV)

Negative-pressure ventilation (NPV) has been used in the fire service in various forms for many years. It consists of ejecting (exhausting) smoke and other airborne contaminants from within a confined space to the outside atmosphere. This involves using some mechanical means of creating airflow, using either a fan (Figure 5.26) or a spray pattern from a nozzle directed out of the confined space through a window, doorway, or other exterior opening (Figure 5.27). The airflow through the

Figure 5.26 An attack team entering beneath a smoke ejector suspended in the doorway.

Figure 5.27 A firefighter performing hydraulic ventilation. *Courtesy of Barry Wagner, IL Fire Service Institute.*

exterior opening creates a mild venturi effect and produces a slight negative pressure (vacuum) within the space. This negative pressure draws cool, fresh replacement air into the space as contaminants are ejected (Figure 5.28). Whenever this or any other form of ventilation is set up, providing adequate openings for replacement air to enter is of critical importance.

A variant of the venturi principle is used in ventilation to create a pressure differential. When a fluid, such as air, is forced under pressure through a restricted orifice, there is a decrease in the pressure exerted against the sides of the constriction and a corresponding increase in the velocity of the fluid. Because the surrounding air is under greater pressure, it rushes into the area of lower pressure (Figure 5.29). This phenomenon can be used to increase the airflow generated by a fan if the fan is properly positioned. Proper placement of the fan is critical when ventilating a structure because its efficiency will be greatly reduced if it is improperly positioned.

When using negative-pressure ventilation, the fan should be placed so that the airstream from the discharge side is centered in an opening with the exhaust flow effectively covering 80 percent to 90 percent of the opening. A pressure drop will be created near the opening as the airstream passes

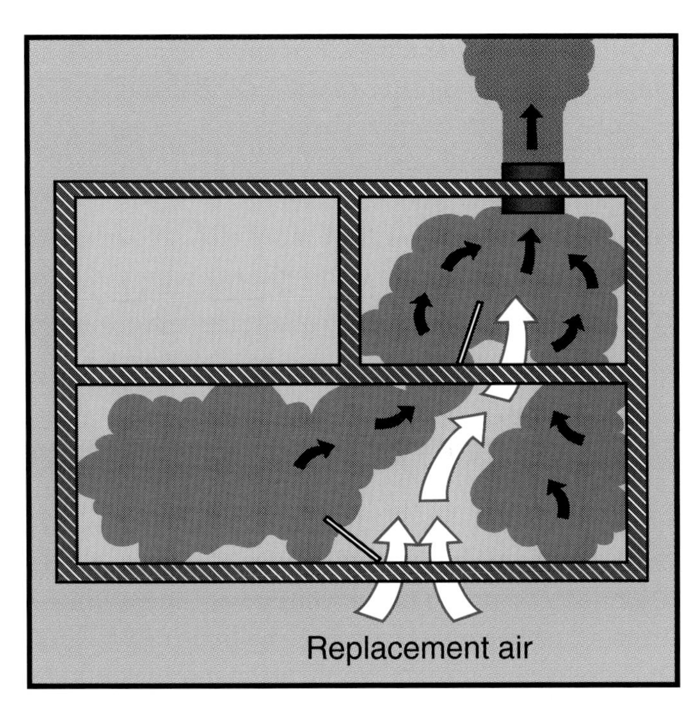

Figure 5.28 Replacement air is drawn into the space during NPV.

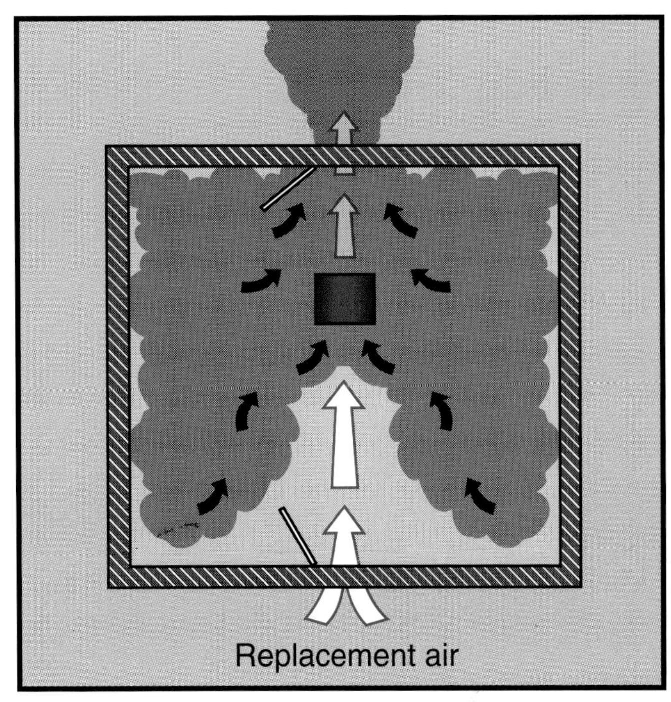

Figure 5.29 The venturi effect created by NPV.

through it. This pressure drop will cause the surrounding air to move toward the airstream and be carried out through the opening. An increased airflow is developed that multiplies the amount of air actually moved by the fan's airstream.

Unlike positive-pressure ventilation, NPV is rarely used as part of the fire attack and is usually started after initial knockdown. Also unlike PPV, NPV is less effective when used with vertical ventilation.

There are two significant disadvantages to NPV. First, the personnel and equipment needed to set up and operate NPV usually have to work within the space being ventilated, subjecting them to the hostile atmosphere they are trying to remove (Figure 5.30). Second, but perhaps most significant, NPV is simply not as effective as PPV in removing smoke or other airborne contaminants from a confined space.

SITUATIONS REQUIRING FORCED VENTILATION

As mentioned earlier, whenever natural ventilation is inadequate or inappropriate or if other elements in the situation limit or preclude natural ventilation, forced ventilation is indicated. Otherwise, there are no definite rules governing when it

Figure 5.30 Firefighters often must work within the hostile atmosphere while setting up NPV equipment.

should or should not be employed. In general, forced ventilation is indicated in the following situations:

- When the type of construction is not conducive to natural ventilation
- When fire is burning below grade in structures or below deck in marine vessels
- When a contaminated atmosphere in nonfire situations must be cleared from a confined space
- When the contaminated area within a confined space is so large that natural ventilation is impractical or too inefficient

Windowless Buildings

In many buildings, both large and small, there may be a shortage of exterior openings (Figure 5.31). These buildings are often of tilt-slab construction, making it extremely difficult and time consuming to breach the exterior walls (Figure 5.32). Other buildings may lack interior passageways or vertical shafts. All of these situations make natural ventilation extremely difficult or very inefficient.

Basement Fires

Basement fires have long been recognized as one of the most difficult and most hazardous types of fires faced by firefighters (Figure 5.33). Very similar to basement fires are fires below deck in ships, barges, and other marine or riverine vessels (Figure 5.34). In each of these cases, access to the fire is limited, and openings for natural ventilation are usually few and of inadequate size. Forced ventilation is usually much more effective in these situations. This topic is covered in greater detail in Chapter 6, High-Rise Structures And Special Situations.

Figure 5.31 A typical windowless building.

Figure 5.32 Typical tilt-stab construction.

Large Interior Areas

When warehouses, factories, or other very large buildings are heavily charged with airborne contaminants, natural ventilation is often inadequate because of a lack of exterior openings (or they are too small compared to the volume of the building's interior) or because there may be little or no wind. In these cases, natural ventilation may need to be supplemented or replaced by forced ventilation.

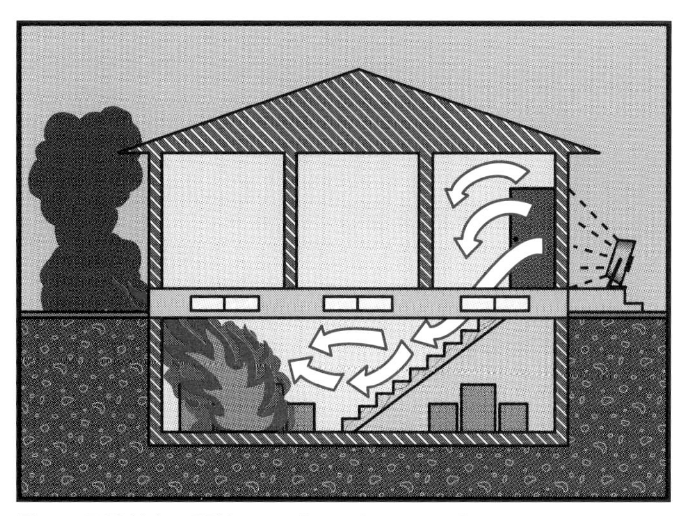

Figure 5.33 Using PPV to ventilate a basement fire.

Figure 5.34 A below-deck fire aboard a typical cargo ship. *Courtesy of Jim Vencil, U.S. Army (Retired).*

Blowers, ejectors, or combinations of the two may be used to rid the building of contaminants (Figure 5.35).

Heavier-Than-Air Gases

In nonfire situations, such as hazardous materials releases in buildings or where heavier-than-air gases have collected in tanks, vaults, tunnels, or open excavations, forced ventilation may be the most efficient and safest way of restoring a tenable atmosphere. Because all liquefied petroleum gases and some nonflammable but toxic gases are heavier than air, below-grade situations can be extremely hazardous, so full protective clothing with SCBA *must* be worn when ventilating these areas. Also, whenever flammable gases or vapors are present, blowers or explosion-proof ejectors must be used, and care must be taken to eliminate sources of ignition in the vicinity before dispersing the contaminants.

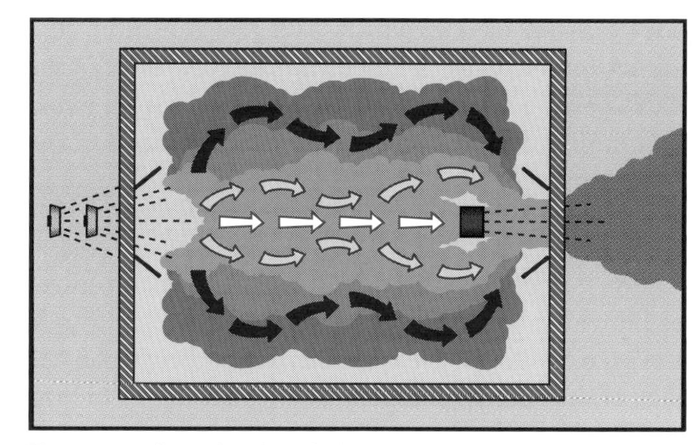

Figure 5.35 A combination of blowers and ejectors can be used to ventilate a large warehouse-type building.

Whenever any below-grade space must be ventilated, as much air movement as possible should be created. This movement may involve using a combination of PPV and NPV by positioning fans at both the entry and the exit openings. This situation may also lend itself to the use of flexible ducting attached to the entry or to the exit fans or to both (Figure 5.36).

WARNING

SCBA *must* be used by all personnel working below grade because oxygen deficiency is likely in these areas, and because numerous toxic gases and vapors are heavier than air and will collect in these areas.

Figure 5.36 Fresh air being supplied in a below-grade operation.

APPLYING FORCED VENTILATION

When applying forced ventilation, firefighters should consider the location and extent of the fire, the location of occupants, and the building construction features in selecting the location for the exit opening. This consideration will help to avoid channeling the fire into uninvolved portions of the building.

Blowers

In setting up PPV at a structure fire, a blower or blowers should be set up about 6 to 8 feet (2 m to 2.4 m) outside the entry point, depending upon the size of the opening and the size and number of fans to be used (Figure 5.37). At the same time, an adequate exit opening through which smoke can escape should be made. The exit opening should be made as close to the seat of the fire as possible and should be opened immediately. The blower or blowers should be started as soon as the first attack crew is ready to enter the building. To establish and maintain a positive pressure within the building, the size of the exit opening must be in proper proportion to the size of the opening where the replacement air is being introduced (Figure 5.38). The exit opening should be about 75 percent to 150 percent as large as the area of the entry opening,

Figure 5.37 Typical positioning of a blower outside of a burning building.

Figure 5.38 Firefighter creating an exit opening of the appropriate size.

depending on the number and size of blowers used. The greater the airflow into the space, the larger the exit opening should be. It is also critically important that blowers — not smoke ejectors — be used to pressurize the building. Smoke ejectors simply do not blow a sufficient volume of air for this purpose.

Just as the fire attack should be made from the unburned side, ventilation air should be introduced from that side also (Figure 5.39). The blower should be set up so that the cone of air created by the fan will completely cover the entry opening to form a seal around the opening (Figure 5.40). The seal can be tested by feeling the direction of airflow around the outside of the doorway with a bare hand. If the cone of air does not seal the entire opening, the fan must be moved back from the opening until a seal is achieved. If the cone extends beyond the corners of the opening, the fan should be moved closer to the opening until a seal is maintained but no airflow is wasted. Additional fans may be added in tandem to multiply the effect. If the fans are of unequal size, the larger fan should be placed about 2 feet (0.6 m) from the entry opening to produce most of the pressurization, and the smaller fan should be placed a few feet (meters) farther back to maintain the seal around the opening (Figure 5.41).

In order for PPV to work, however, it is imperative that firefighters do not arbitrarily open windows as they progress through the building

Figure 5.39 Ventilation air being supplied from the unburned side of the building.

Figure 5.40 The cone of air created by the blower should completely cover and form a seal around the entry opening.

Figure 5.41 Typical positioning of fans used in tandem.

during the primary search (a standard operating procedure in some departments) unless the interior doors are closed after each room is searched. If firefighters arrive to find numerous windows already open, they must realize that it may not be possible to establish effective PPV without closing some of the interior doors.

For the positive pressure to be maintained, the area of the exit opening or openings *must* be kept in proper proportion to the area of the entry opening and the volume of replacement air being introduced (Figure 5.42). If the exit opening is too small, smoke will not be cleared efficiently, thereby limiting visibility and increasing smoke damage. This problem can be recognized by the smoke remaining suspended or moving toward the exit opening very slowly. If the exit opening is too large, the positive pressure cannot be maintained, and the effort will be ineffective. This problem can be recognized by rapid smoke movement near the entry opening but a slowing down to little or no movement toward the exit point.

If a large, compartmentalized building is charged with smoke, individual rooms may be ventilated by closing the doors to the other rooms and creating a vent opening in the room to be ventilated

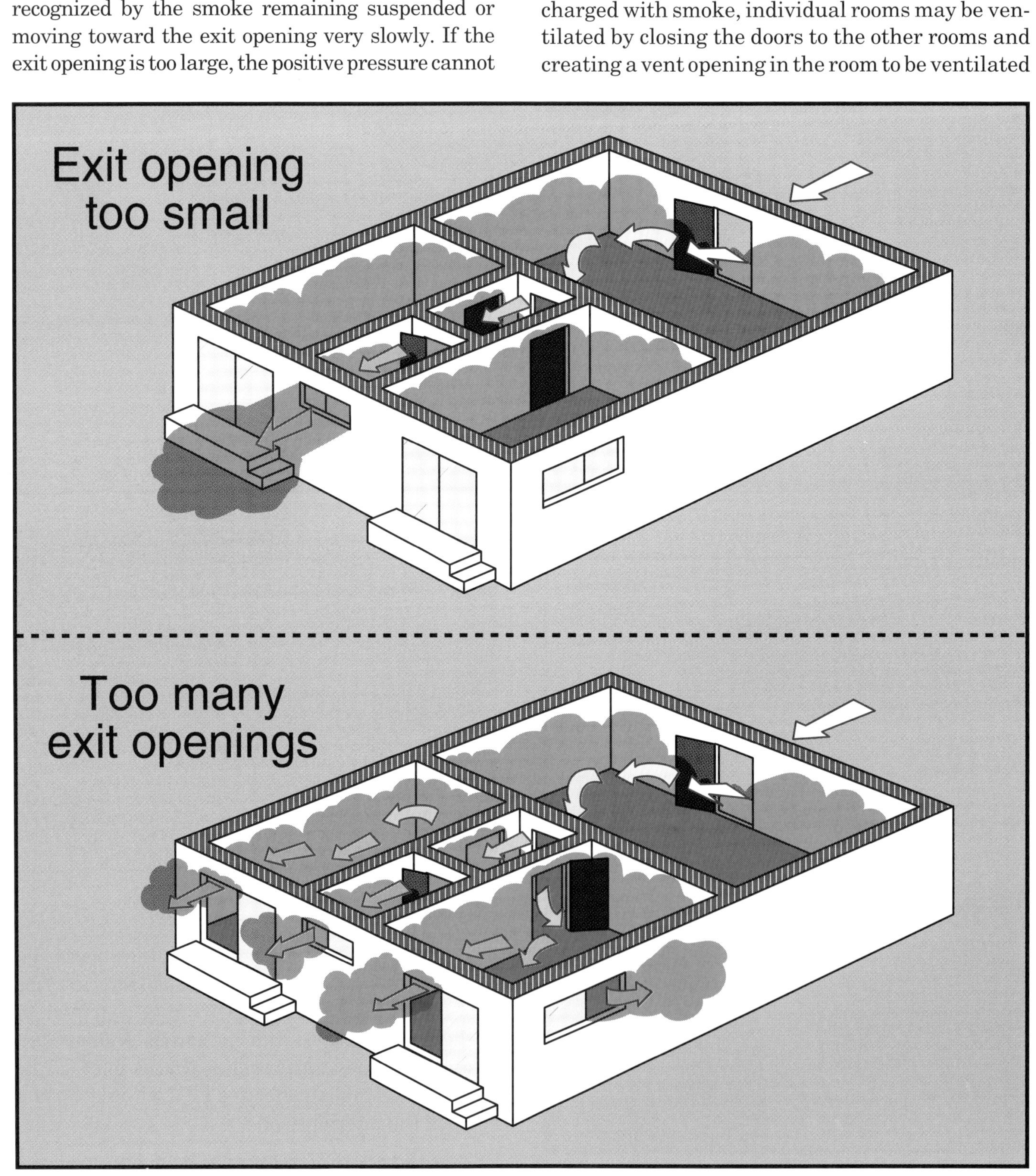

Figure 5.42 The area of the exit openings and entry openings must be kept in proper proportion.

(Figure 5.43). By progressing from room to room, keeping all rooms closed except the one being ventilated, an entire floor or an entire building can be systematically cleared of smoke very quickly. To reduce smoke damage, the room(s) most heavily charged with smoke should be cleared first.

If spaces within a sealed or windowless building must be ventilated, a slightly different technique must be used in addition to the PPV in the building as a whole. With PPV in operation in the building, individual rooms may be purged by placing an electric fan at the floor level in the open doorway to the room being ventilated (Figure 5.44). Air is forced into the room at the floor level, displacing smoke accumulated near the ceiling and forcing it out at the top of the open doorway. As the smoke exits the room, it is then carried out of the building by the PPV.

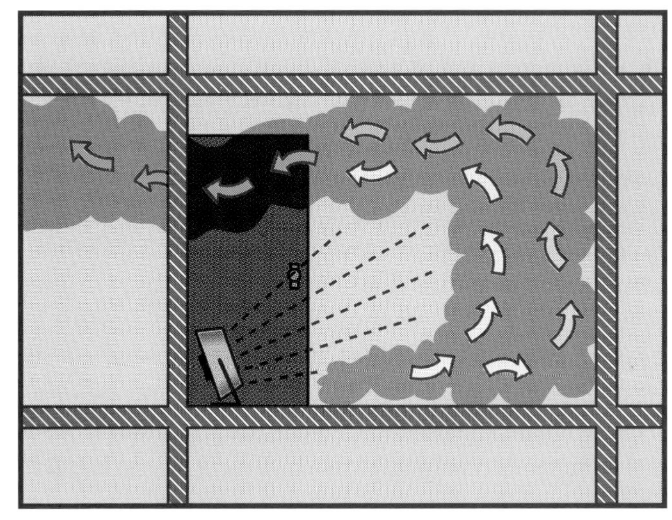

Figure 5.44 An electric fan being used to purge an interior room of smoke.

Unlike negative-pressure ventilation, PPV can remove airborne contaminants from areas far removed from the fans being used without having to

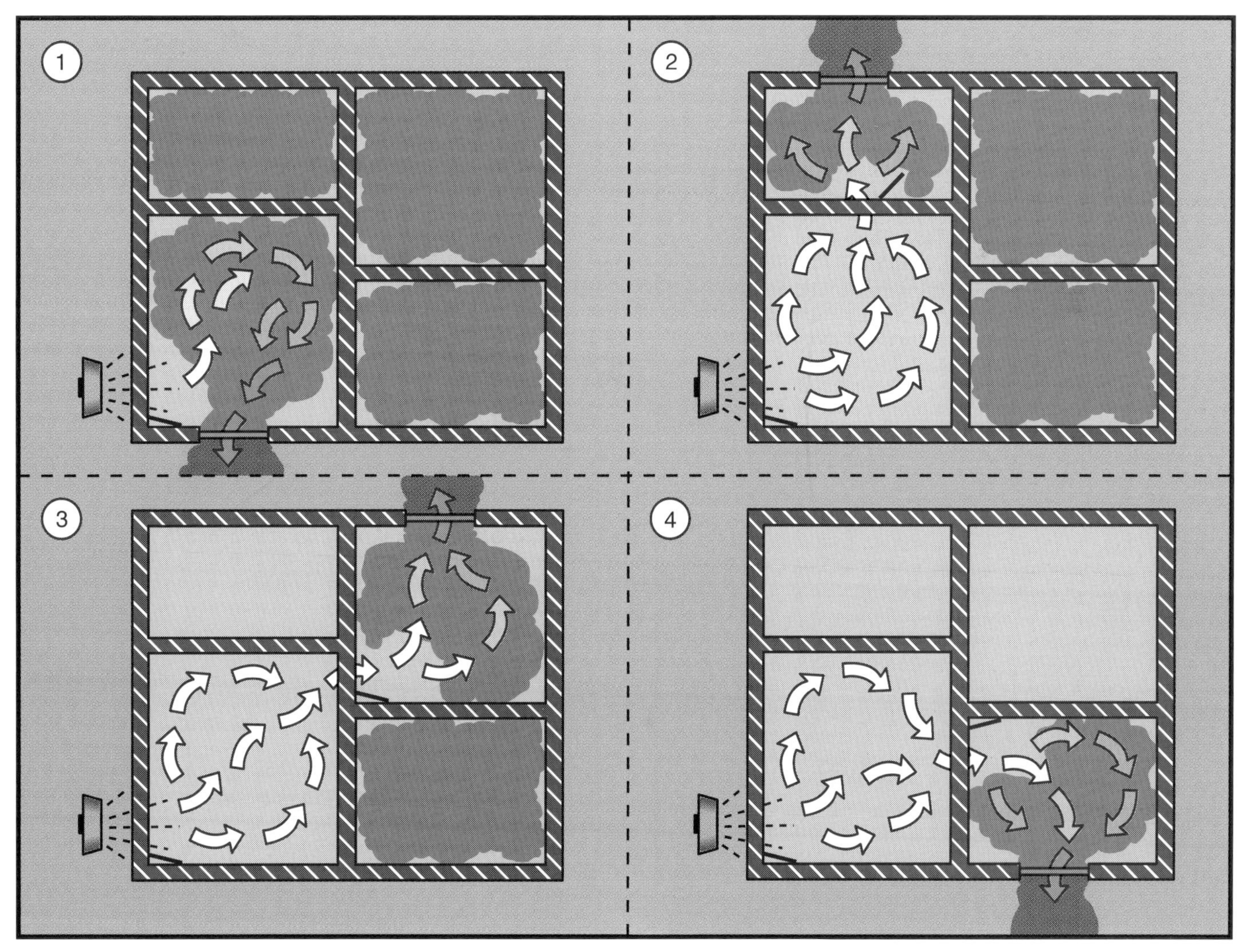

Figure 5.43 Rooms being ventilated by opening windows and systematically opening and closing interior doors.

use flexible ducting. This is especially important in ventilating upper floors of high-rise buildings. In this instance, blowers set up on the ground level can force smoke out of rooms that are more than twenty floors up in the building (Figure 5.45). At some point above that level, the effect of the ground-level blowers must be supplemented by additional fans on or near the floors to be ventilated (Figure 5.46). Otherwise, except for high-rise buildings beyond the reach of aerial ladders, most — if not all — of the work required to set up PPV can be done from outside the building. This means that personnel assigned to ventilate the building do not have to endure working in the same conditions they are trying to eliminate.

Once knockdown has been achieved, PPV should be continued into the overhaul phase of the operation (Figure 5.47). This continuation will accomplish a number of things. First, it will hasten the elimination of products of combustion from the interior of the building and replace the contaminants with cooler air. This process will speed the restoration of a tenable atmosphere. It will also aid in the detection of hidden hot spots by supplying additional air so that they will begin to produce visible smoke sooner.

The usual sequence at a typical structure fire is to apply PPV during and as part of the initial fire attack. After initial knockdown of the fire, attack personnel may be relieved and sent to Rehab, but the PPV should be continued. The atmosphere within the structure should be sampled by someone wearing SCBA (Figure 5.48). All personnel working in the contaminated atmosphere must continue

Figure 5.45 A blower set up on the ground level can force smoke out of rooms that are more than twenty floors up.

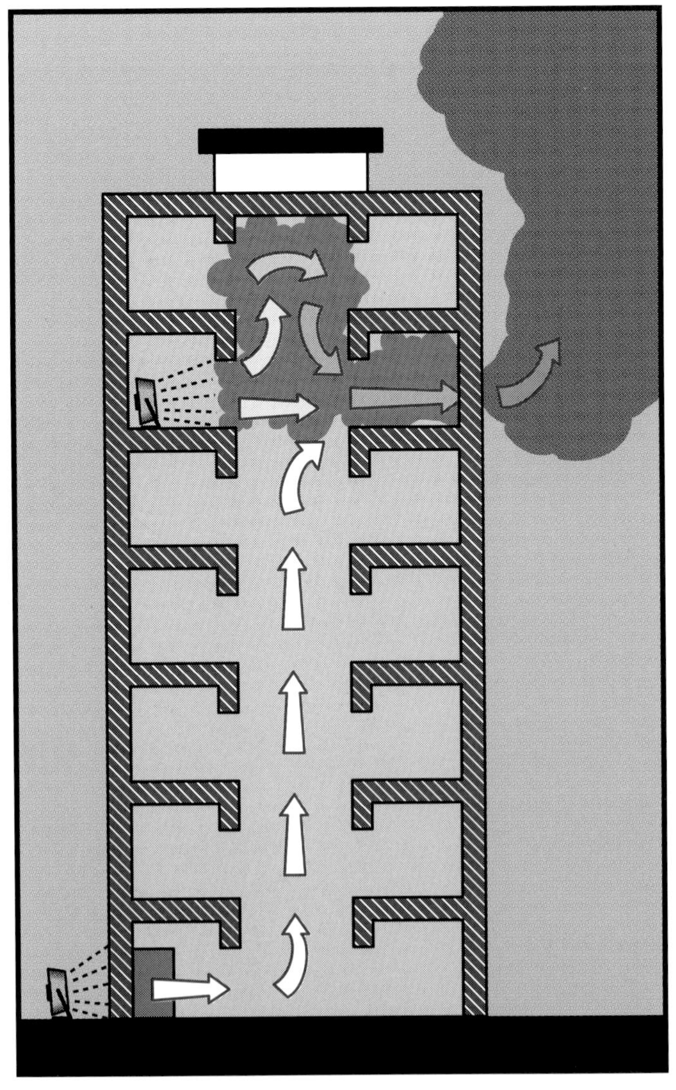

Figure 5.46 An additional fan set on or near the floor to be ventilated to supplement the ground-level blowers.

to wear SCBA as long as the carbon monoxide (CO) level exceeds the established standard. If the other gases present are known, then it is recommended that their levels be monitored also. See Chapter 1 for more information on using CO as an index gas.

Figure 5.47 PPV being continued as attack team personnel take a break.

Figure 5.48 A firefighter using a CO meter to sample the atmosphere within a building.

Smoke Ejectors

Smoke ejectors have been the mainstay of fire service ventilation for many years. While they have their limitations, they can still be used to good effect if applied to the correct situation in the correct manner.

Smoke ejectors are designed for and are most often used in exit openings for NPV. They are usually set up in windows or doorways, using built-in C-clamps (Figure 5.49) or adjustable hangers to position them at the top of the opening (Figure 5.50). To be most effective in these positions, the exit opening surrounding the ejector must be sealed in some way to prevent churning and the pulling of

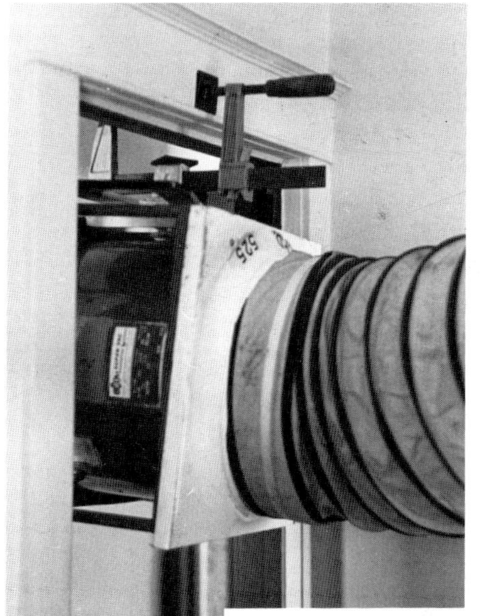

Figure 5.49 A smoke ejector suspended in a doorway using a built-in C-clamp. *Courtesy of Lisle-Woodridge Fire District.*

Figure 5.50 A smoke ejector suspended from an adjustable hanger.

contaminants back into the space. Salvage covers, hall runners, or specially designed drapes are often used for this purpose, but this covering blocks the opening and limits its use for access (Figure 5.51). Sometimes smoke ejectors are hung from tops of doors that have been blocked open. This arrangement makes sealing the area around the ejector very difficult and may create a safety hazard for firefighters who must pass through these openings (Figure 5.52).

Airflow tests have shown that the optimum position for a smoke ejector is inside the space to be ventilated, about 6 feet (2 m) back from the exit opening. The ejector should be placed anywhere from the floor level up to about 4 feet (1.2 m) above the floor, depending on whether the contaminants to be ejected are lighter or heavier than air (Figure 5.53). The critical point is to position the fan so that maximum airflow is achieved through the exit opening. This point can only be determined by moving the fan both closer to and farther away from the exit opening and watching the amount of smoke being ejected. The unit should be positioned at the point where the maximum amount of smoke

Figure 5.52 It is difficult to seal the area around a smoke ejector that is hung from the top of a door.

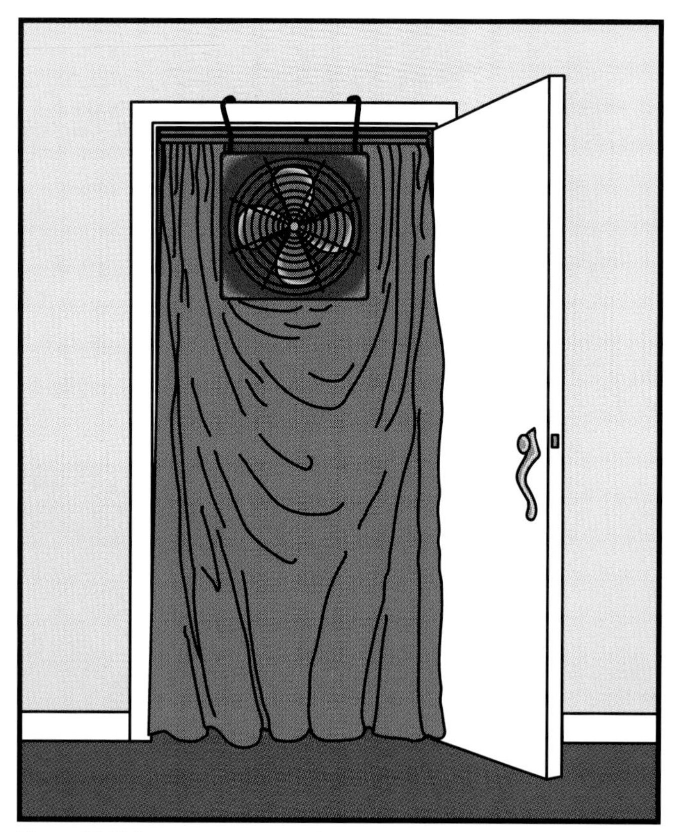

Figure 5.51 The open area around a smoke ejector sealed with a salvage cover.

Figure 5.53 A firefighter placing a smoke ejector in the optimum position.

appears to be exhausted and where no churning or recirculation occurs (Figure 5.54).

Some unusual situations, such as smoke accumulations in remote, windowless spaces, often below grade, can force firefighters to become very creative in positioning smoke ejectors for maximum effect. Such positioning can involve using ladders, pike poles, salvage covers, or other equipment (Figure 5.55).

If it is necessary or desirable to remove contaminants from an interior space to the outside atmosphere without contaminating the intervening spaces, flexible ducting can be attached to a smoke ejector (Figure 5.56). This technique can also be used effectively in some below-grade situations where airflow is unusually restricted.

Using smoke ejectors in tandem, so that one ejector moves air toward another ejector that either

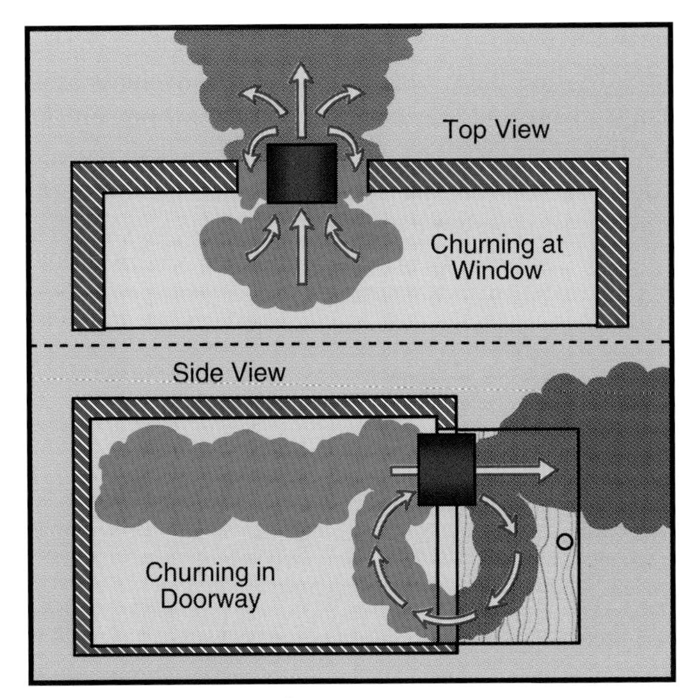

Figure 5.54 How churning often occurs.

Figure 5.55 A few of the ways to position smoke ejectors.

moves it to yet another ejector or exhausts it to the outside, has been used successfully in some unusual situations (Figure 5.57). While this method is not the most efficient use of these fans, some situations leave firefighters few options, and some air movement is better than none.

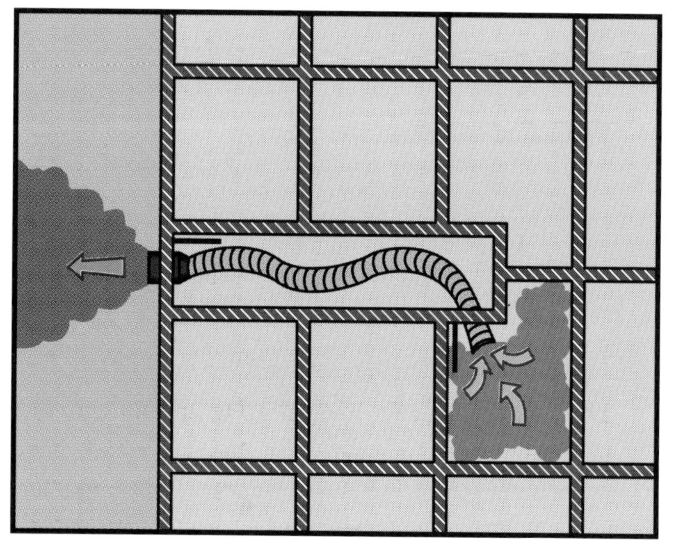

Figure 5.56 Flexible ducting attached to a smoke ejector.

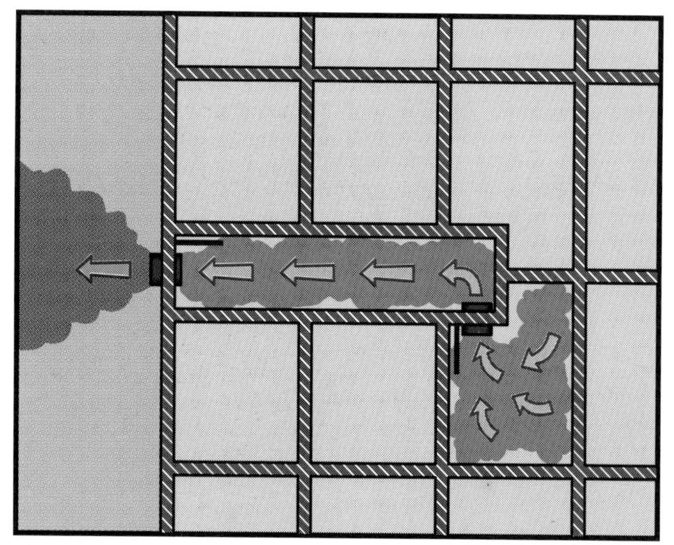

Figure 5.57 Smoke ejectors can be used in tandem.

Hydraulic Ventilation

Whenever a nozzle on a charged line is opened, the fog or spray stream forces a certain amount of air ahead of the pattern, and replacement air is drawn in behind the nozzle. A fog stream or spray stream directed through a window or doorway will draw large quantities of heat and smoke out the opening through which the stream is directed (Figure 5.58). The use of fog or spray streams in forced ventilation is a proven technique, and the degree of effectiveness depends upon how, where, and when the stream is applied.

There are, however, some disadvantages of using hydraulic ventilation. There may be an increase in the amount of water damage within the structure, there will be some additional impact on the available water supply, and in periods of freezing temperatures there may be an increase in the buildup of ice in the area surrounding the building. Another disadvantage of using fog or spray streams for ventilation is that the person operating the nozzle usually must remain inside the space being ventilated. This situation not only subjects that person to the effects of the hostile environment, but the operation may have to be stopped each time the operator runs out of breathing air.

Compared with smoke ejectors, properly applied fog or spray streams have been found to remove two to four times more smoke, depending on the type and size of the nozzle, the angle of the spray pattern, and the position of the nozzle in relation to the exit opening. A stream directed

Figure 5.58 A fog stream directed through a window to draw heat and smoke out. *Courtesy of Barry Wagner, IL Fire Service Institute.*

through the opening with a 30-degree to 60-degree fog pattern covering 85 percent to 90 percent of the opening has been found to provide the best results for ventilation. To apply this technique with a handline, a firefighter in full protective gear and SCBA should extend the nozzle through the exit opening, open the nozzle, and adjust the fog or spray pattern. The nozzle should then be retracted through the opening until the maximum air movement is achieved. This maximum will usually occur when the nozzle is positioned about 2 feet (0.6 m) inside the exit opening (Figure 5.59).

Figure 5.59 The pattern recommended for effective hydraulic ventilation.

Larger openings permit greater airflow, so using a doorway can sometimes be more effective than using a window. If large openings, such as roll-up doors, are available, large volumes of air can be moved using a master stream set up in the same way a handline nozzle is used to eject smoke from a structure (Figure 5.60). However, regardless of the size of the exit opening, the fog pattern should not be increased beyond a 60-degree angle in an attempt to cover more of the opening. If it is increased beyond this angle, the stream begins to lose air-moving efficiency. Instead, the nozzle should be moved closer to or farther from the exit opening to maintain the maximum airflow through the opening. If the fog pattern still will not cover the opening sufficiently, more than one nozzle will have to be used or a larger nozzle employed.

This technique can be applied to both small and large openings. As mentioned before, extremely large structures, such as warehouses, can be vented through roll-up door openings using portable monitors. A portable monitor setup using the same guidelines as those used for a handline nozzle will quickly move large volumes of air. Unlike a handline nozzle, however, a master stream appliance cannot easily be moved forward or backward to adjust the amount of draft being created. Once in operation, the water flow must be shut down if the appliance

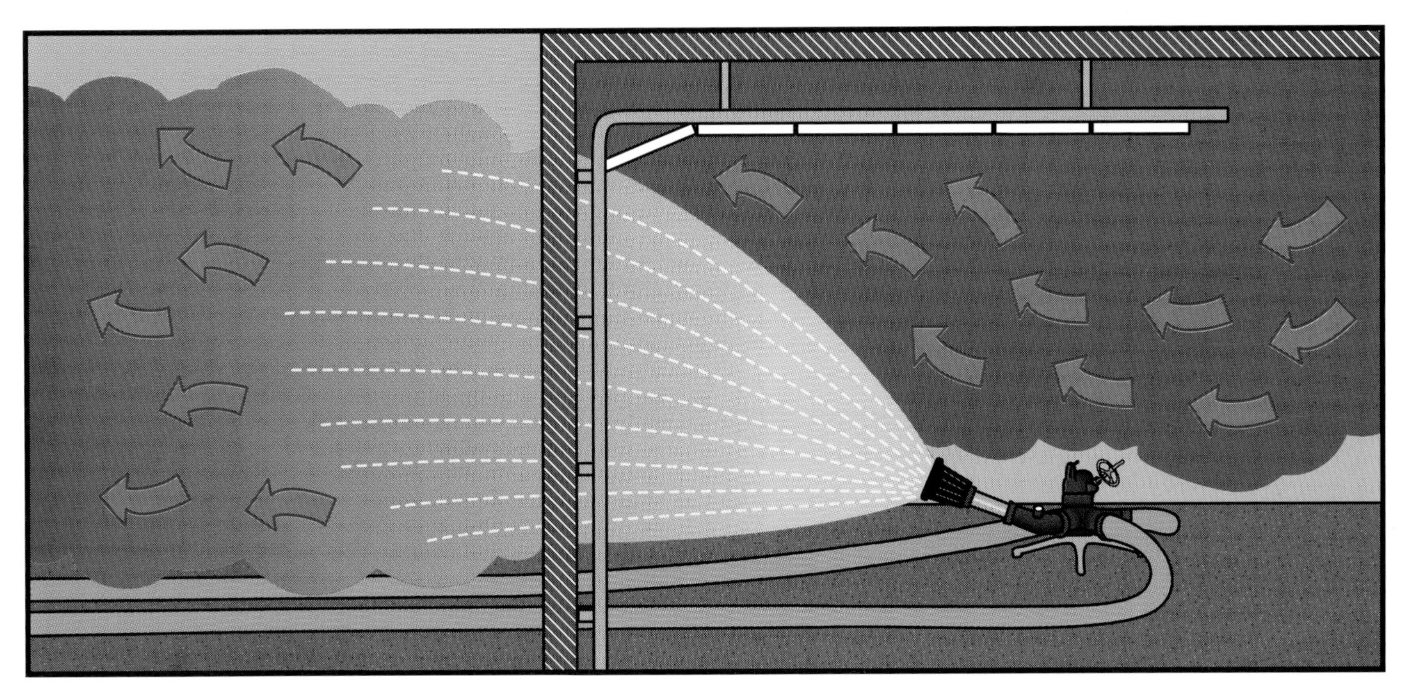

Figure 5.60 A master stream set up for hydraulic ventilation.

must be moved. Obviously, using this method will require a substantial water supply.

Fixed Systems

In most cases, residential and light-duty commercial air-handling systems offer firefighters little assistance in the removal of smoke and other airborne contaminants. However, in some postfire or nonfire situations, such as when there is only airborne contamination in the living space, the system can assist in clearing the atmosphere if it is designed to draw intake air from outside and introduce it into the building. In other situations, attic fans may be used to exhaust cold smoke or other contaminants from the living space. The attic scuttle can be opened, thus allowing the attic fan to pull the contaminants through the attic and exhaust them to the outside (Figure 5.61). For more information on using fixed systems, see Chapter 6, High-Rise Structures And Special Situations.

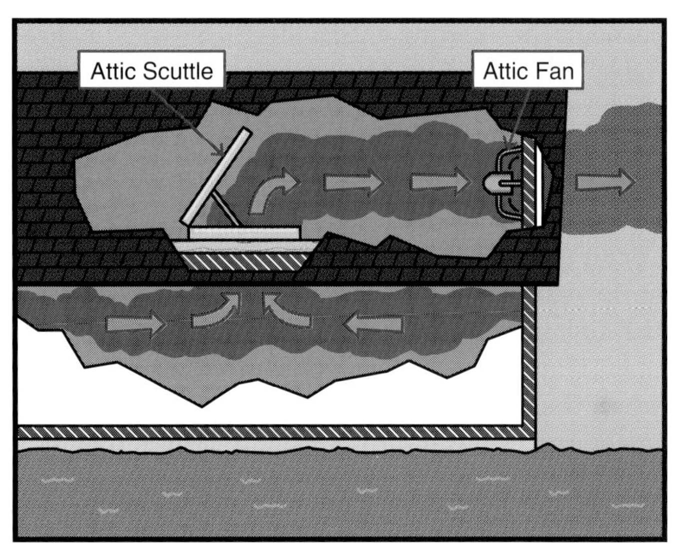

Figure 5.61 One technique for exhausting cold smoke after a fire.

Chapter 5 Review

Directions

The following activities are designed to help you comprehend and apply the information in Chapter 5 of **Fire Service Ventilation**, Seventh Edition. To receive the maximum learning experience from these activities, it is recommended that you use the following procedure:

1. Read the chapter, underlining or highlighting important terms, topics, and subject matter. Study the photographs and illustrations, and read the captions under each.
2. Review the list of vocabulary words to ensure that you know the chapter-related meaning of each. If you are unsure of the meaning of a vocabulary word, look the word up in the IFSTA **Orientation and Terminology** glossary or a dictionary, and then study its context in the chapter.
3. On a separate sheet of paper, complete all assigned or selected application and review activities before checking your answers.
4. After you have finished, check your answers against those on the pages referenced in parentheses.
5. Correct any incorrect answers, and review material that was answered incorrectly.

Vocabulary

Be sure that you know the chapter-related meanings of the following words:

- access *(144)*
- adverse *(147)*
- ambient *(141)*
- arbitrarily *(152)*
- auxiliary *(142)*
- churning *(157)*
- compound *(142)*
- conducive *(150)*
- constriction *(149)*
- critical *(149)*
- damper *(146)*
- ducted *(144)*
- ejecting *(139)*
- enhanced *(140)*
- entrainment *(139)*
- erratic *(139)*
- exhaust *(139)*
- impede *(143)*
- imperative *(152)*
- intake *(144)*
- in tandem *(152)*
- integral *(148)*
- integrated *(146)*
- marine *(149)*
- misnomer *(139)*
- obscure *(141)*
- preclude *(149)*
- proportion *(152)*
- purge *(155)*
- riverine *(150)*
- vacuum *(139)*
- velocity *(149)*

Application Of Knowledge

1. Describe your department's SOPs for setting up PPV at a structure fire. If your department has no SOPs, get together with a classmate, and list in order those procedures you feel the department should standardize. *(Local protocol)*
2. Describe your department's SOPs for setting up smoke ejectors for NPV at a structure fire. If your department has no SOPs, get together with a classmate, and list in order those procedures you feel the department should standardize. *(Local protocol)*

Review Activities

1. Define each of the following terms: *(139)*
   - positive pressure
   - negative pressure
   - venturi effect
2. List eight advantages of forced ventilation. *(140)*
3. List five disadvantages of forced ventilation. *(143)*
4. Compare and contrast the functions and uses of smoke ejectors and blowers. *(144, 145)*
5. Briefly explain the operation of smoke ejectors. *(144)*
6. Briefly explain the operation of blowers. *(144, 145)*
7. Explain how hoseline nozzles can be used for forced ventilation. *(145)*
8. Discuss the effectiveness of residential and commercial climate control systems in terms of forced ventilation. *(146)*
9. Discuss the pros and cons of PPV and NPV, weighing the advantages and disadvantages of each. *(147-159)*
10. Describe situations for which forced ventilation would be appropriate. *(150)*
11. List nonfire situations for which forced ventilation would be appropriate. *(151)*
12. List advantages and disadvantages of hydraulic ventilation. *(160)*

6

High Rise And Special Situations

This chapter provides information that addresses the following performance objectives of NFPA 1001, *Standard for Fire Fighter Professional Qualifications* (1992):

Chapter 3 — Fire Fighter I

3-9.3 Describe the advantages and disadvantages of the following types of ventilation:
- (a) Vertical
- (b) Horizontal
- (c) Trench/strip
- (d) Mechanical
- (e) Mechanical pressurization
- (f) Hydraulic

3-9.10 Define procedures for the types of ventilation referred to in section 3-9.3.

Chapter 4 — Fire Fighter II

4-9.1 Identify the manual and automatic venting devices found within structures.

4-9.2 Describe the operations and considerations necessary to control the spread of smoke and fire through duct systems, including:
- (a) Determining location and routing of ducts
- (b) Shutting down systems to prevent spread of heat and smoke
- (c) Examining duct system after thorough ventilation
- (d) Checking false ceilings or framing enclosing duct systems
- (e) Checking duct system outlets
- (f) Determining if duct system has openings, smoke dampers, or smoke detectors

4-9.4 Identify the location of the opening, the method to be used, and the precautions to be taken when ventilating a basement.

Safety Points

In its discussion of high-rise structures and special situations, this chapter addresses the following safety points:

- Neither firefighters nor building occupants should use an elevator during a high-rise fire unless the officer in charge on the fire floor(s) determines that it is safe to do so.
- Only firefighters who have been trained to do so should be allowed to use a Billy Pugh net or similar device to get to the roof of a building via helicopter.
- Firefighters working in any stairwell or other vertical opening must be aware of the potential dangers of the stack effect in tall buildings.
- The Safety Officer should establish a clear zone of at least 200 feet (60 m) around a burning high-rise building to protect those on the street from falling debris.
- If a ventilation opening must be created at the top of an atrium in a high-rise building, firefighters must be *extremely* careful because a fall from that point is likely to be fatal.

Chapter 6
High-Rise Structures And Special Situations

INTRODUCTION

With the urbanization of many formerly rural or residential areas, the cost of land has risen significantly. As a result, builders have found it more cost effective to build up, rather than out. This practice has resulted in high-rise buildings being erected even in small communities, presenting a new fire fighting challenge for many fire departments (Figure 6.1). A department that has previously had to deal with only one- and two-story buildings may now be faced with having to perform rescues or fight fires above the reach of its longest ladders.

Figure 6.1 A lone high-rise building in a small community.

The same economic factors that have created the need for high-rise buildings have also spurred the development of windowless and underground structures (Figure 6.2). To deal with these extraordinary structures, fire department personnel must develop new strategies based upon the construction and facilities of the building and the available resources. This chapter deals with the problems associated with ventilating these unusual structures.

Figure 6.2 A typical windowless building.

FIRE FIGHTING IN HIGH-RISE BUILDINGS

High-rise fires present numerous challenges for firefighters. Limited access, large numbers of offices or apartments, masses of occupants, falling glass and other debris, smoke and fire spread through vertical shafts, locked interior doors, low water pressure, and crews having to climb an extraordinary number of stairs are some examples of what firefighters can expect when fighting a high-rise fire (Figure 6.3). Crews may have to deal with stranded occupants, unfamiliar floor layouts, complex elevator and ventilation systems, built-in

Figure 6.3 A working high-rise fire.

fire protection equipment, and difficult communications, each of which may complicate fire fighting operations.

The staffing required for operations in this type of building may be four to six times as great as that required for fires in lower structures. The problems of communication and coordination between attack and ventilation crews can be compounded as the number of involved personnel increases. In addition to the extra personnel required to handle the logistics of fighting a high-rise fire, the physical exertion required to merely reach the fire floor may mean that attack crews will have less energy left to fight the fire. Consequently, attack crews may have to be relieved sooner than they normally would, and crews may have to be rotated more frequently (Figure 6.4).

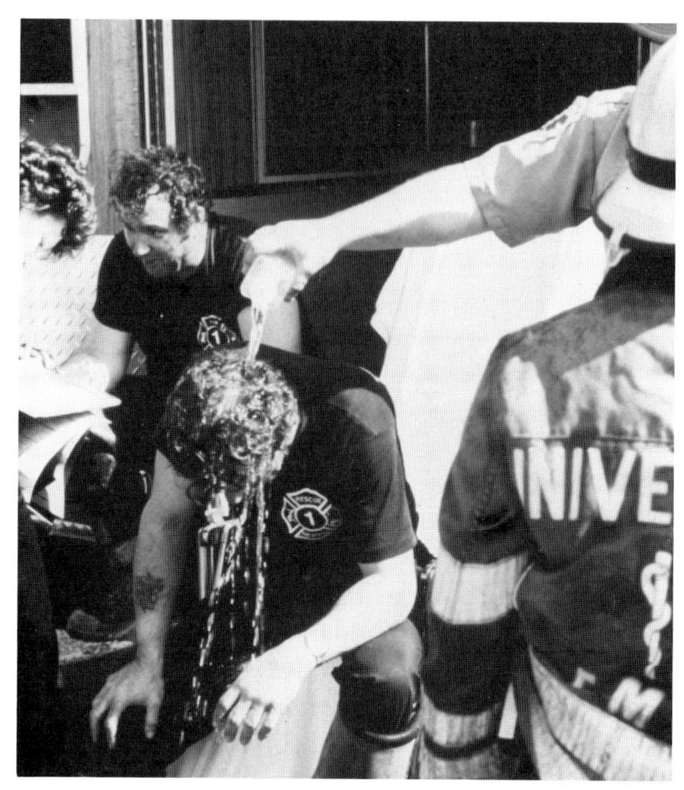

Figure 6.4 An attack crew cooling off in Rehab. *Courtesy of Ron Jeffers.*

Most fire departments mount their attack from the floor below the fire, and relief crews and spare equipment are concentrated in *Staging*, two floors below the fire floor. During a high-rise fire, firefighters should never enter the building empty-handed. If they do not have to carry ventilation tools and equipment, they should carry spare SCBA cylinders, hose, tools, lights, etc., to Staging (Figure 6.5).

High-rise buildings often have a variety of elevators including low-rise, high-rise, and freight elevators. Low-rise elevators are those that serve the lower floors of the building, with high-rise elevators serving the upper floors. For example, some elevators may serve floors 1 through 10, others 10 through 20, and so on (Figure 6.6). In this manual, however, for an elevator to be considered "low- rise," it must not serve the fire floor(s). There may also be one or more express elevators that run nonstop from the street level to the top floor. Whenever the fire situation allows, using freight elevators to transport firefighters and additional equipment can be advantageous because these elevators are designed to carry large, heavy loads. Several fully equipped firefighters carrying additional

Figure 6.5 A firefighter carrying spare SCBA cylinders to Staging.

equipment will be less likely to overload freight elevators than passenger elevators. However, freight elevators should only be used when authorized by the officer in charge on the fire floor(s).

WARNING

Elevators that serve the fire floor or above should not be used by occupants or firefighters unless the officer in charge on the fire floor has determined that it is safe to do so.

The elevator car may be automatically called to the fire floor due to an elevator control malfunction,

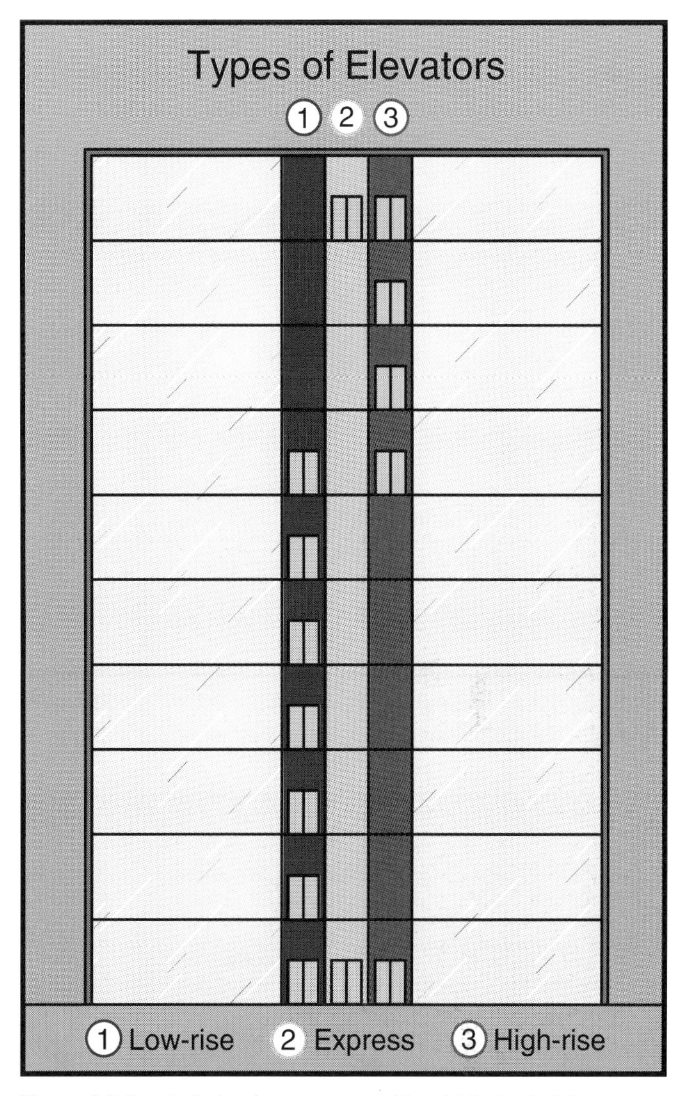

Figure 6.6 A typical elevator arrangement in a high-rise building.

or a power failure may strand an elevator car above the fire or between floors. Therefore, those assigned to *Lobby Control* should bring all elevators to the ground floor and lock them there (Figure 6.7). Low-rise elevators (those that do not serve the fire floor or above) equipped with manual fire department control systems may then be used for evacuating occupants and for shuttling firefighters and equipment partway to the fire floor.

The mushrooming effect that commonly occurs on the top floors of smaller buildings does not occur in the same way in very tall buildings if insufficient heat is generated to move the smoke and fire gases to the top of the building. Heat dissipates into the environment as smoke and fire gases travel through the building. These products of combustion will rise through any vertical opening until they encounter an obstruction or until their temperature is reduced to the temperature of the surrounding air. When this equalization of temperature occurs, the smoke and fire gases lose their buoyancy, cease to rise, and *stratify* — forming layers or clouds of smoke within the building (Figure 6.8). This process can occur several floors below the top floor of the building. The products of combustion will then spread laterally and downward until the building is ventilated.

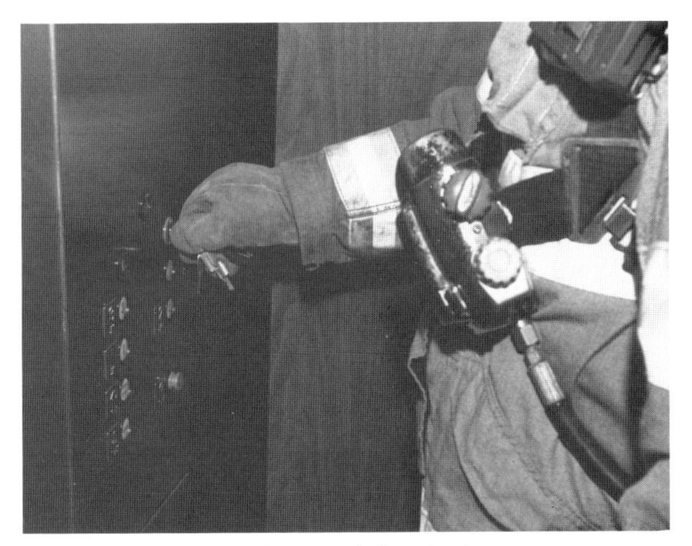

Figure 6.7 A firefighter taking control of an elevator.

Stratification of smoke and fire gases can create a highly toxic atmosphere many floors above the fire, even where there is little heat. Ventilating this cooled, stratified smoke out of the building can be accomplished by using positive-pressure blowers to create a controlled flow of air up the stairwell and horizontally across the smoke-filled floors (Figure 6.9).

HIGH-RISE VENTILATION

High-rise buildings may be ventilated in several ways. Vertical (top) ventilation, horizontal

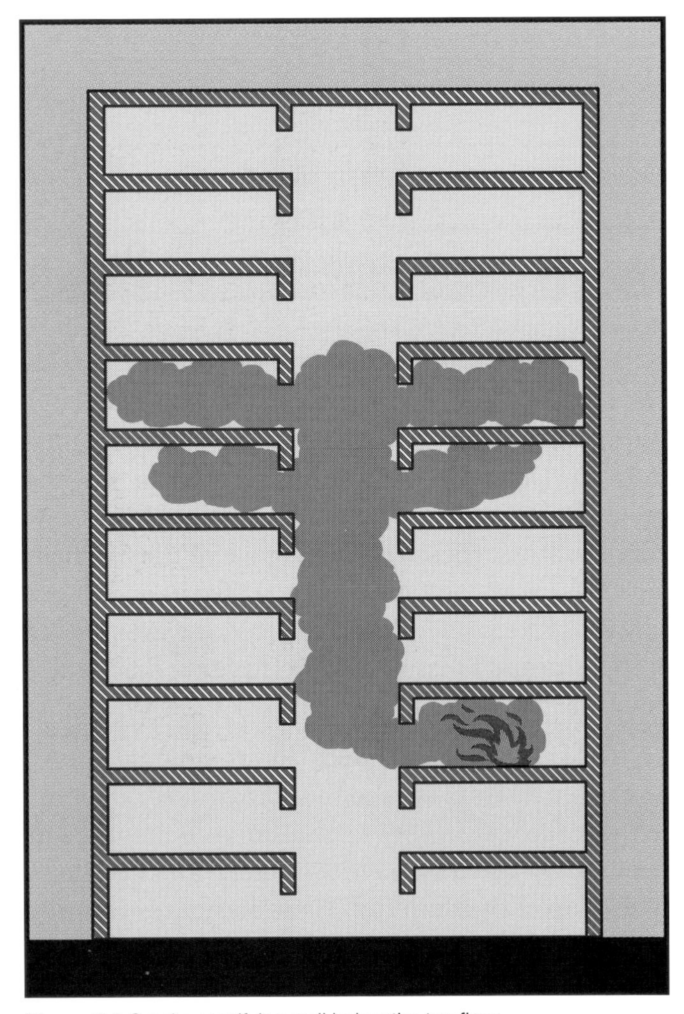

Figure 6.8 Smoke stratifying well below the top floor.

Figure 6.9 A positive-pressure blower at ground level ventilating upper floors.

ventilation of the fire floor, and horizontal ventilation above and below the fire floor are the options available at a high-rise fire. Because of the numerous factors affecting smoke movement in a high-rise fire, it may be necessary to employ more than one of these methods. The following discussion of airflow within high-rise buildings relates primarily to *natural* ventilation. The natural airflow within a high-rise building can be enhanced or even reversed through mechanical means, especially when *positive-pressure* ventilation is used. See Chapter 5, Forced Ventilation, for more information on mechanical or forced ventilation.

In high-rise buildings, fire and smoke may spread rapidly through pipe shafts, stairways, elevator shafts, air-handling systems, and other vertical openings . In some instances, ventilation must be accomplished horizontally on the fire floor or on the floors immediately above the fire where smoke and heat may have spread. Despite the danger of the fire lapping into floors above the fire (Figure 6.10) and the hazards of shards of broken glass falling onto those in the street below, horizontal ventilation may be the most appropriate method in some situations — such as when ventilating vertically would endanger occupants attempting to leave the building. In other situations, the fire can and should be vented vertically through stairwells or other vertical shafts, taking advantage of the so-called "stack effect" in which heat and smoke naturally rise through these openings.

Figure 6.10 Fire lapping into the floors above. *Courtesy of Los Angeles City Fire Department.*

CAUTION: Elevator shafts should not be used for ventilation because of the danger created by leaving shaft doors open when visibility may be reduced by smoke or darkness.

Top (Vertical) Ventilation

During pre-incident planning, top ventilation of serious fires in modern high rises should be considered because top ventilation can prevent or reduce mushrooming on the upper floors and does not promote lapping, which is always a danger when venting horizontally. A pre-incident inspection of the roof will reveal the existence of any roof vents that may lend themselves to the fire ventilation process or of any automatic smoke vents that may reduce or eliminate the need for additional top ventilation by firefighters (Figure 6.11).

Figure 6.11 A typical automatic smoke vent.

GETTING PERSONNEL TO THE ROOF

One of the biggest challenges in top venting high-rise buildings is getting the vent group and their equipment to the roof. The roofs of most high-rise and even some low-rise buildings are beyond the reach of aerial devices, so some alternative means of getting to the roof must be found. The various ways of reaching the roof of a high-rise building involve using an aerial device whenever possible, using an interior stairway, using low-rise elevators and the interior stairway, and using helicopters. Each of these methods has certain advantages and disadvantages.

Aerial apparatus. When aerial devices are able to reach the roof of a high-rise building, using them is the preferred method of getting to the roof because it is the fastest, safest, and most direct route. Firefighters using an aerial device will not have to contend with the congestion of personnel and equipment that often develops in the interior stairways of high-rise buildings during a fire. In the majority of cases, however, the aerial device will not reach the roof but can be used as far as it will reach to access an exterior fire escape or to force entry into an upper floor of the building. The vent group can then continue to the roof via the interior stairway.

The interior stairway. While most high-rise buildings have more than one interior stairway, few have more than one that penetrates the roof and would be useful for ventilation purposes. References to *the* interior stairway assumes that the stairway penetrates the roof. Pre-incident inspection of the building should reveal which stairway penetrates the roof, whether it is a smokeproof tower, whether it can be pressurized from below, and whether it is likely to be used for access during fire attack. During a fire, the interior stairway may be relatively free of smoke but can be heavily congested with occupants leaving upper floors, with firefighters carrying equipment to Staging, and with charged attack lines. Vent crews should coordinate with attack crews before deciding which interior stairway to use.

Low-rise elevators. As stated earlier, any elevator that serves the fire floor(s) or above should *NOT* be used by occupants or firefighters unless the officer in charge *on the fire floor* determines that it is safe to do so. Low-rise elevators that do not reach the fire floor may be used to shuttle firefighters and equipment to the highest floor served by the elevator (Figure 6.12). The vent group can then use

Figure 6.12 Firefighters entering an elevator that has been determined to be safe to use.

an interior stairway to reach their destination, but they should assess heat and smoke conditions before committing themselves to using the stairway.

Helicopters. One of the most direct means of moving personnel and equipment to the roof of a high-rise building — and of removing occupants stranded on the roof — is through the use of helicopters, weather and smoke conditions permitting. To use helicopters safely and most effectively during a fire requires close coordination between fire suppression personnel and the helicopter crew. This level of coordination can only be achieved through pre-incident planning and realistic training between the two groups.

Procedures need to be developed for identifying safe landing zones, for transporting fire and rescue personnel to the roof in Billy Pugh nets or similar devices (Figure 6.13), and for removing occupants from the roof, some of whom may be injured. Once the procedures have been developed, they need to be tested in realistic exercises and revised as necessary. To maintain an adequate level of readiness, truck crews and helicopter crews need to train together on a regular, periodic basis.

Figure 6.13 Firefighters being transported in a Billy Pugh net. *Courtesy of Dallas / Fort Worth International Airport.*

CHANNELING THE SMOKE

In buildings having only one stairwell that penetrates the roof, this natural "chimney" may be used to ventilate smoke, heat, and fire gases from various floors. The bulkhead door on the roof must be blocked open or removed from its hinges before the stairway doors on the fire floors are opened (Figure 6.14). Removal of the bulkhead door ensures that it cannot be closed accidentally, upsetting the established ventilation process and allowing the shaft to become filled with superheated gases. During fire fighting operations, firefighters should enter this natural chimney above the fire only when they are certain that it is safe to do so.

Timing and coordination with rescue and fire suppression operations is extremely important (Figure 6.15). Ventilation directed up the one stairwell that penetrates the roof must be delayed until all occupants above the fire floor are either evacuated or moved to an area of refuge within the building. Firefighters should also be in positions of safety prior to executing the ventilation order. Once the ventilation operation has begun, the stairway may be untenable, even for firefighters in full protective clothing. For these reasons, the stairway used for ventilation should not be used for fire attack.

Figure 6.14 A rooftop bulkhead with the door removed.

Figure 6.15 Vent team leader coordinating with rescue and attack teams.

STACK EFFECT

Natural top ventilation of high-rise buildings depends entirely on the stack effect (Figure 6.16). The intensity of the stack effect will depend on how airtight the outside and inside walls are, the distance between the upper and lower openings, and the difference between the temperature inside the building and the temperature outside. The greater the distance between the upper and lower openings and the greater the difference in temperature, the more intense the stack effect. If a high-rise building is vented at the roof and at the street level, the direction and intensity of airflow within the building will depend primarily on the relative temperature differences. If it is hotter inside than outside, the airflow will be inward at the bottom and outward at the top. But if the outside air is hotter than the air inside, the flow will be reversed (Figure 6.17). If there is little or no distance between the upper and lower openings and if inside and outside temperatures are equal, no natural airflow will take place.

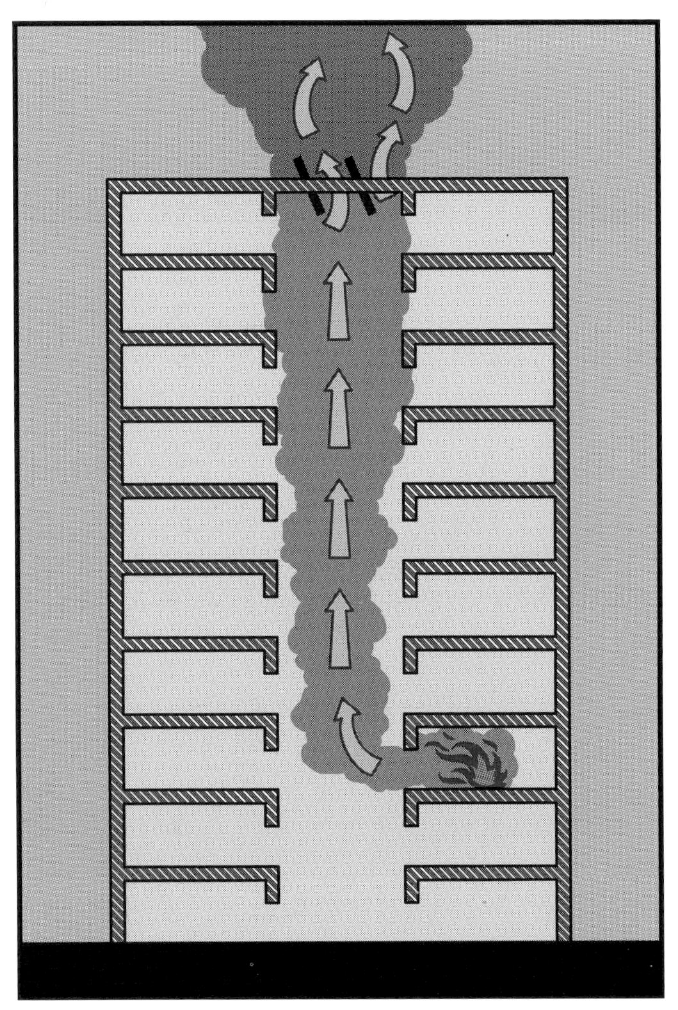

Figure 6.16 Natural top-ventilation of high-rise buildings depends entirely on the stack effect.

Figure 6.17 Stack effect is reversed if the outside air is hotter than inside.

Effects Of Wind

Wind can raise or lower the neutral pressure plane within a building and can have a major influence on high-rise ventilation. Wind will produce a positive pressure on the windward side of the building (which would tend to raise the neutral pressure plane) and a negative pressure on the leeward side of the building (which would tend to lower the neutral pressure plane) (Figure 6.18). Before venting the fire floor horizontally, it must be determined that air movement will be conducive to effective ventilation. For example, depending upon where the seat of the fire is located, ventilating on the lee side of the building can make best use of the negative pressure condition created by the wind, but ventilating on the windward side could work against effective ventilation and spread the fire into uninvolved areas (Figure 6.19).

Wind will also influence the stack effect by raising or lowering the neutral pressure plane. Positive wind pressure will cause the neutral pressure plane to rise, and negative pressure will cause it to fall. Ventilating horizontally below the neutral pressure plane will draw air into the building, thus spreading the smoke throughout the interior rather than ejecting it to the outside. On the other hand, ventilating above the neutral pressure plane will allow the smoke to escape to the outside (Figure 6.20). The closer that ventilation takes place to the neutral pressure plane, the less positive or negative effects it will exert.

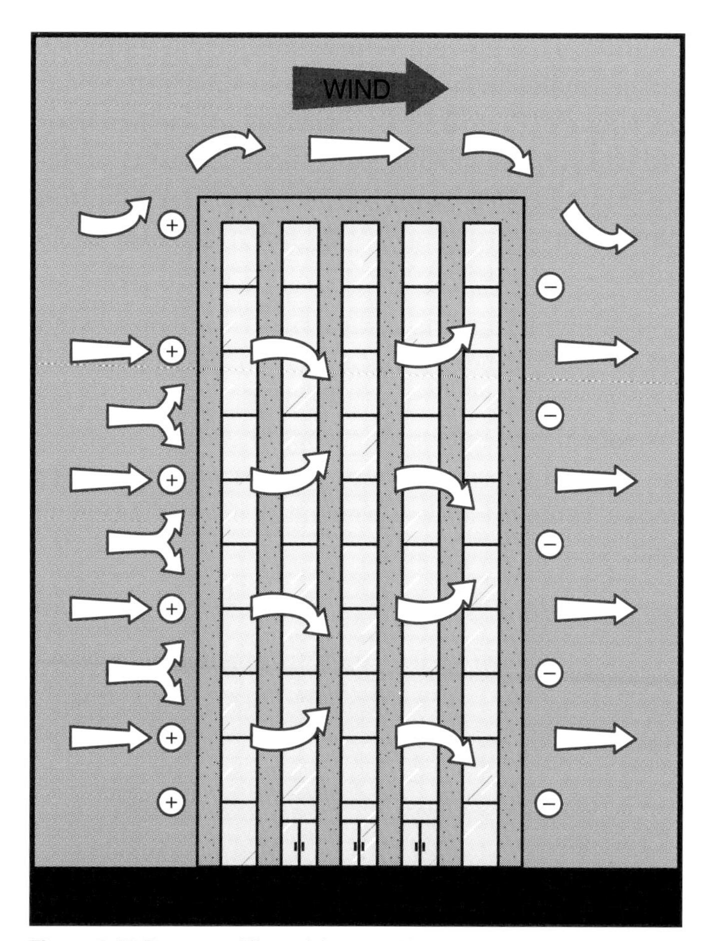

Figure 6.18 Pressure differential created by the wind.

Figure 6.19 Possible effects of wind on horizontal ventilation.

Ventilating Below The Fire

Ventilating below the fire floor is not a common practice but may be necessary when smoke has spread to the floors below the fire due to negative stack conditions or mushrooming. The most common technique is to vent these floors horizontally; the results achieved can be enhanced by pressurizing the entire building with blowers.

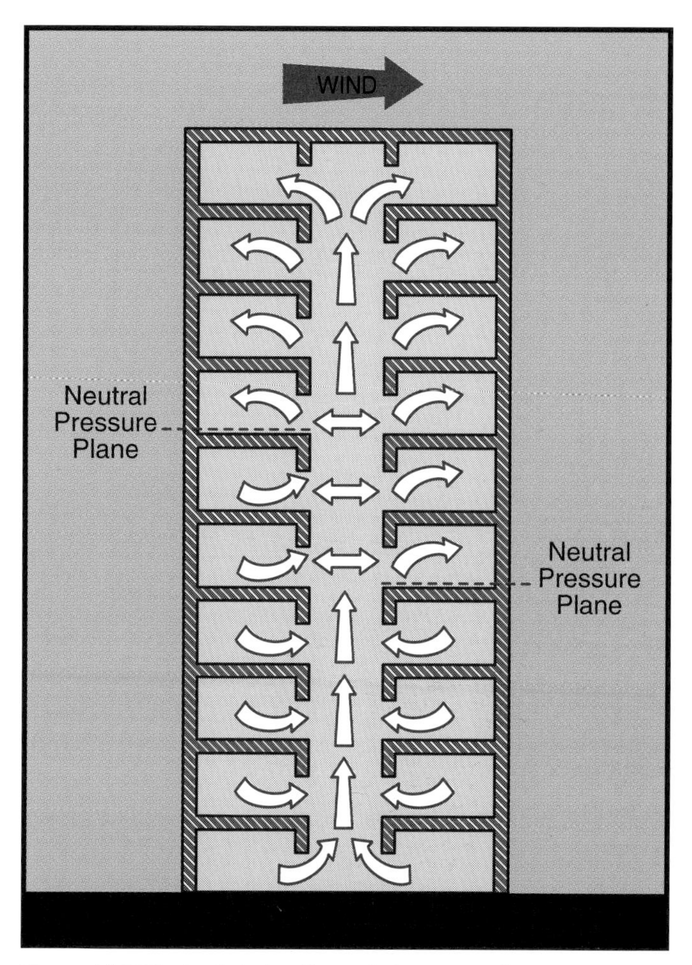

Figure 6.20 Effects of wind on the neutral pressure plane.

Ventilating The Fire Floor

Prior knowledge of the layout of the building is extremely important. Attempting to ventilate through a stairwell that does not penetrate the roof or through a dead-end corridor can seriously delay the extinguishment of the fire and increase fire and smoke damage. Because many modern high rises are sealed buildings, the windows may not be designed to be opened (Figure 6.21); therefore, horizontal ventilation may necessitate breaking windows. Because this procedure is difficult, time consuming, and potentially quite dangerous to those in the street below, horizontal ventilation of the fire floor should be attempted only when top ventilation is impractical. Mechanically ventilating up a stairwell, across the smoke-filled fire floor, and out through the roof via another stairwell is much preferred in sealed buildings (Figure 6.22).

Figure 6.21 A typical modern high-rise building with windows that are not designed to be opened.

Figure 6.22 An effective method of ventilating a sealed building.

Ventilating Above The Fire

Ventilating above the fire floor will be most effective if the process is started at the top of the building. This procedure provides a clear exit path for the smoke and gases when doors on the fire floor are opened (Figure 6.23). Starting at the fire floor and working upward is less efficient, may contribute to increased fire spread and smoke damage, and may place ventilation personnel in greater jeopardy.

Following extinguishment, when only cold smoke is left to be evacuated, venting above the fire floor eliminates the need for breaking windows for horizontal ventilation. Also, opening up floors near the neutral pressure plane only minimally affects smoke removal unless mechanical ventilation is used.

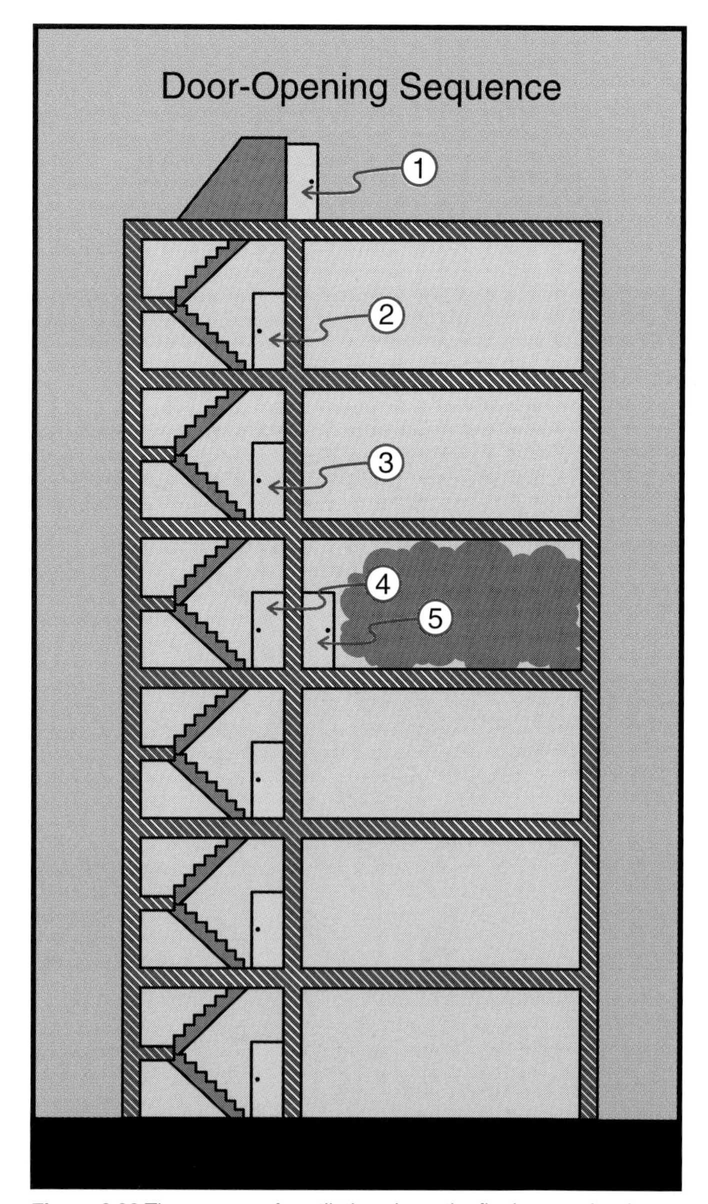

Figure 6.23 The process of ventilating above the fire is started at the top to provide a clear exit path for smoke and gases.

HVAC AND SMOKE-CONTROL SYSTEMS

Traditionally, engineers designed HVAC systems considering only the demands for controlling the environment of high-rise buildings under normal conditions. Little thought was given to how such systems might affect a fire or to how they might contribute to the spread of smoke and fire throughout a building. More recently, however, energy costs have forced improvements in the efficiency of these systems. For example, most now contain controllable dampers that can be selectively opened and closed. With these features, HVAC systems are more easily adapted to also serve as smoke-control systems.

Correct and effective use of an HVAC system can limit the spread of smoke and fire gases, improve operating conditions for fire fighting personnel, and increase the likelihood of survival for building occupants. Although the actual manipulation of a building's HVAC system for smoke control should be left to the building engineer, firefighters should have an understanding of the HVAC system's capabilities and limitations (Figure 6.24). This knowledge will help them work more effectively with the engineer to use the system to the best advantage.

Some heavy-duty HVAC systems can be of considerable assistance in removing cold smoke following a fire. These systems are quite varied in design and complexity, so as a general rule, firefighters should not attempt to manually control these systems. When problems develop in buildings having these systems, the HVAC system should be shut down until the location and extent of the fire is known and the building's maintenance engineer or other knowledgeable personnel can be called to the scene. However, because elevators, stairwell pressurization fans, and other vital equipment require electricity to operate, the building's *utilities* should not be shut down *arbitrarily* when a fire is burning within the building (Figure 6.25).

To effectively use an HVAC system to control smoke movement, fire department personnel should observe the following guidelines:

- The HVAC system should be operated by a qualified building engineer, not by firefighters.
- The HVAC system may be used to assist in locating the origin of the fire.
- The HVAC system should be used to limit the extension of fire and smoke to the smallest possible area.
- The HVAC system should not promote the growth or extension of fire or smoke beyond the area of origin.

Figure 6.24 Firefighters should learn an HVAC system's capabilities and limitations.

Figure 6.25 A firefighter awaiting an order to shut down a building's utilities.

- The HVAC system should provide fresh, uncontaminated air to any occupants who are trapped or are located in a designated safe refuge area within the building.

The HVAC system should have smoke detectors in the ducts to shut down the system in the event smoke enters the ducts. If not, smoke from a fire in a high-rise may spread to several floors before the HVAC system can be shut down manually. To accomplish a manual shutdown with the least possible delay, contact must be made with the building engineer as soon as fire department units arrive. Under the direction of the incident commander, the building engineer may be able to use the HVAC system for exhausting smoke from the building.

The building engineer should be consulted during pre-incident planning to allow firefighters an opportunity to learn the capabilities and limitations of the HVAC system. This planning will also provide fire department personnel an opportunity to inform the building engineer that during a fire he or she would be expected to remain available to the IC for consultation and for actual manipulation of the HVAC system under fire department direction.

BUILT-IN VENTILATION DEVICES

Roof vents and curtain boards are most common in large buildings having wide, unbroken expanses of floor space. Some industrial or warehouse facilities are so large that their floor space is measured in acres. If these structures are not properly vented and/or protected by other permanently installed systems for limiting fire spread, the entire contents are vulnerable to smoke and fire damage.

Roof vents and curtain boards have proven to be effective in limiting the spread of fire, releasing heated fire gases, and reducing smoke damage. NFPA 204M, *Guide for Smoke and Heat Venting*, the standard that provides guidelines for the design and installation of smoke and heat venting equipment, recommends using automatic heat-activated roof vents and curtain boards. The following is general information on various types of vents and curtain boards; however, firefighters need to become familiar with the specific types in use in their local areas.

Automatic Roof Vents

Automatic roof vents are intended to limit the spread of fire within a building by releasing heat and smoke to the outside before the fire mushrooms throughout the building. Because they work automatically, these vents may eliminate the need for additional ventilation by fire department personnel.

Automatic roof vents take advantage of the fact that fire gases tend to rise, so they are placed at the highest point of the roof (Figure 6.26). Although some are now activated by smoke detectors, most still operate through the use of fusible links connected to spring-loaded or counterweighted cover assemblies. When the temperature of the fusible link reaches its designed fusing temperature, the link separates, allowing the vent covers to open (Figure 6.27). Automatic locking devices help to ensure that the covers remain open, even in gusty winds.

CAUTION: In darkness or heavy smoke, firefighters should be extremely careful when working on roofs with automatic vents to avoid falling into them.

Figure 6.26 A typical automatic roof vent.

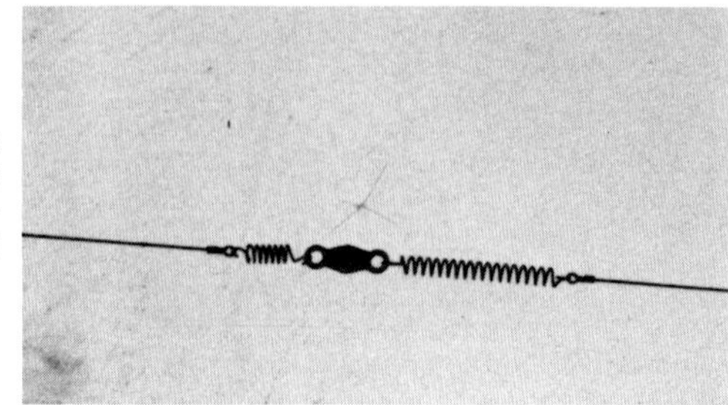

Figure 6.27 A typical fusible link keeping an automatic roof vent closed.

Heat-activated roof vents may not open automatically when sprinkler heads discharge near them because the sprinklers may prevent the fire from developing enough heat to activate the vents. Attempts to forcibly open automatic roof vents from the exterior can do extensive damage to their operating mechanisms and may be dangerous to firefighters because of their spring-loaded opening systems. Therefore, firefighters should become familiar with the manual rclease mechanisms of automatic roof vents in their areas (Figure 6.28).

Atrium Vents

A growing number of high-rise hotels and office buildings are being constructed with an atrium in the center of the structure. These large, vertical openings lend themselves to the stack effect, so building codes in most areas require that they be equipped with automatic vents. These automatic vents are usually designed to be activated by either smoke or heat (Figure 6.29). The consequences of a firefighter falling into an open vent over an atrium in a high-rise building are obvious, so the earlier caution about working on roofs in darkness and/or heavy smoke is especially important in these situations.

Monitors

Monitor vents are usually rectangular structures that penetrate the roofs of single-story buildings (Figure 6.30), but they may be found on high-rise buildings as well. The monitor may have metal, glass, wired glass, or louvered sides. Those with glass sides rely upon the glass breaking to provide ventilation in case of a fire. If the fire has not yet generated enough heat to break the glass, the glass will have to be broken by firefighters. Monitors with solid walls should have at least two opposite sides hinged at the bottom and held closed at the top with a fusible link that will allow them to open in case of fire (Figure 6.31).

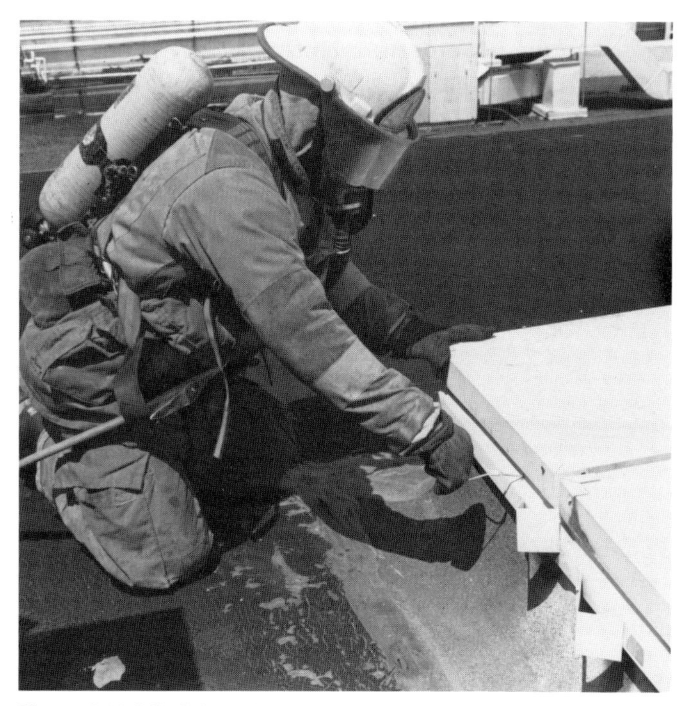

Figure 6.28 A firefighter about to manually open an automatic roof vent.

Figure 6.29 Most building codes require that an atrium be equipped with automatic vents that can be activated by either smoke or heat. *Courtesy of Bilco Company, New Haven, CT.*

Skylights

As described in Chapter 4, Vertical Ventilation, skylights may be used to ventilate heat and smoke in the event of fire. Skylights equipped with thermoplastic panels or ordinary window glass can act as automatic vents because the temperature of a fire will melt the plastic or cause the glass to break and fall (Figure 6.32). In the absence of skylights with thermoplastic panels or those with automatic

Figure 6.30 A typical rectangular monitor vent on the roof of a single-story building.

Figure 6.31 A roof monitor with one side open.

Figure 6.32 A typical skylight equipped with thermoplastic panels.

venting, firefighters will have to remove the skylight or break the glass panes to ventilate the building. However, skylights equipped with wired glass usually must be removed for ventilation purposes.

Curtain Boards

Curtain boards, also known as *draft curtains*, are fire-resistive half-walls that extend down from the underside of the roof. They generally extend a distance equal to at least 20 percent of the vertical distance from the floor to the roof but not lower than 10 feet (3 m) above the floor. The areas encompassed by curtain boards will generally be those containing critical industrial processes and/ or concentrations of flammable liquids or other hazardous materials with high fire potential.

The function of curtain boards is to limit the horizontal spread of heat and smoke by confining it to a relatively small area directly over its source. They also concentrate heat and smoke directly under automatic roof vents to accelerate their activation. Curtain boards may also accelerate the activation of automatic sprinklers in the area, and this process will help to get water onto the fire sooner. However, as mentioned earlier, it may also slow or prevent the activation of automatic roof vents. If the roof vents do not open automatically, firefighters will have to open them manually.

UNDERGROUND STRUCTURES

Underground structures, such as basements, utility vaults, tunnels, etc., rarely provide an opportunity for effective natural ventilation and usually require mechanical ventilation (Figure 6.33). In addition to the difficulties often encountered in gaining access for ventilation below grade, there is a much greater likelihood of having to work in a hostile atmosphere.

WARNING

All personnel working below grade *must* wear SCBA because of the likelihood of oxygen deficiency in addition to whatever toxic gases may be present.

Because basements are the most common underground structure, this section deals primarily with ventilating them. Most basements and subbasements have openings such as stairways, elevators, chutes, or windows. However, opening them for ventilation may be a slow, difficult process

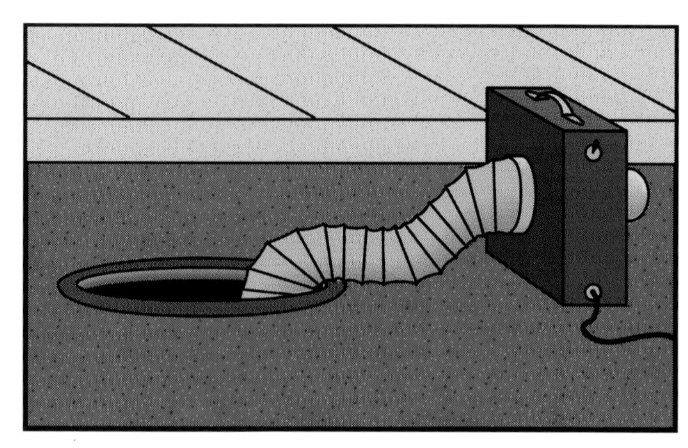

Figure 6.33 A manhole being mechanically ventilated.

because the openings may be blocked or secured at the street level by iron gratings, steel shutters, wooden doors, or some combination of these. Building features, such as stairways, elevator shafts, pipe chases, laundry chutes, air-handling systems, and other vertical openings, may contribute to the spread of fire and smoke from the basement to upper floors. Therefore, early and effective ventilation of basement fires is critically important to successful fireground operations.

To prevent or reduce the upward spread of smoke and fire from an involved basement, it must be ventilated as quickly as possible. If an opening of adequate size can be made opposite the point of entry into the basement, the fastest way to ventilate the basement is to set up positive-pressure blowers at the point of entry (Figure 6.34). This procedure will allow the vent group to stay out of the basement while still ventilating it effectively; however, the rest of the structure must be closely and constantly monitored for signs of fire extension through walls or other vertical channels.

Figure 6.34 Using a positive-pressure blower to ventilate a basement.

WINDOWLESS BUILDINGS

Many modern buildings have few, if any, windows (Figure 6.35). They depend on skylights and artificial lighting for interior illumination. Some windowless buildings have various forms of glass walls that admit light but that are of little value for ventilation purposes (Figure 6.36). These walls may appear to be fragile, but they can be as resistant to being breached as most masonry walls.

Figure 6.35 A typical modern windowless building.

Figure 6.36 A typical wall of glass blocks.

The absence of exterior openings severely limits the opportunities for horizontal ventilation and increases the likelihood of backdraft conditions developing within these structures. Both of these problems underscore the need for fast and effective vertical ventilation of fires in windowless buildings.

Chapter 6 Review

Directions

The following activities are designed to help you comprehend and apply the information in Chapter 6 of **Fire Service Ventilation**, Seventh Edition. To receive the maximum learning experience from these activities, it is recommended that you use the following procedure:

1. Read the chapter, underlining or highlighting important terms, topics, and subject matter. Study the photographs and illustrations, and read the captions under each.
2. Review the list of vocabulary words to ensure that you know the chapter-related meaning of each. If you are unsure of the meaning of a vocabulary word, look the word up in the IFSTA **Orientation and Terminology** glossary or a dictionary, and then study its context in the chapter.
3. On a separate sheet of paper, complete all assigned or selected application and review activities before checking your answers.
4. After you have finished, check your answers against those on the pages referenced in parentheses.
5. Correct any incorrect answers, and review material that was answered incorrectly.

Vocabulary

Be sure that you know the chapter-related meanings of the following words:

- arbitrarily *(177)*
- atrium *(166)*
- conducive *(174)*
- congestion *(172)*
- designated *(178)*
- dissipate *(169)*
- encompass *(180)*
- enhance *(171)*
- generated *(169)*
- jeopardy *(176)*
- lapping *(171)*
- manipulation *(177)*
- shard *(171)*
- shuttling *(169)*
- spurred (verb) *(167)*
- stratify *(170)*
- urbanization *(167)*
- vulnerable *(178)*

Application Of Knowledge

1. Take an informal tour of a public high-rise structure in your area — a hospital, library, office building, etc. Jot down ventilation ideas, anticipated problems in ventilating, and any permanent features that would affect ventilation efforts. Compare your notes with your department's pre-incident plans for the structure toured. *(Local protocol)*

2 Read three or four pre-incident plans for high-rise buildings in your jurisdiction. Based on the information the plans contain, outline anticipated ventilation measures you might take were a fire to occur on an upper floor of any one of these structures. *(Local protocol)*

Review Activities

1. Identify the following terms associated with high-rise and special-situation ventilation:
   - attack floor *(168)*
   - Billy Pugh net *(172)*
   - curtain board *(178)*
   - neutral pressure plane *(174)*
   - pipe chase *(181)*
   - roof monitor *(179)*
   - stack effect *(171)*
   - staging floor *(168)*
2. List the challenges presented by high-rise fires. *(167, 168)*
3. Briefly describe each of the following types of elevators:
   - low-rise *(168)*
   - high-rise *(168)*
   - freight *(168, 169)*
   - express *(168)*
4. Discuss some of the hazards presented by elevator use during a high-rise fire — both to the occupants and to the fire fighting crew. *(168, 169)*
5. Explain how smoke stratification occurs in a high-rise fire. *(169, 170)*
6. List the three ventilation techniques commonly used in high-rise fires. *(170, 171)*
7. List advantages and disadvantages of top (vertical) ventilation in high-rise structures. *(171-176)*
8. Discuss the pros and cons of each of the following ways of getting the ventilation crew and their equipment on the roof for vertical ventilation operations. *(172)*
   - aerial apparatus
   - interior stairway
   - low-rise elevator
   - helicopter
9. List safety measures that should be taken when channeling smoke, heat, and fire gases up a stairwell that penetrates the roof. *(173)*
10. Explain how airtight walls, temperature differentials, and the distance between ventilation openings affect the stack effect when performing natural vertical ventilation of high-rise buildings. *(173)*

11. Describe how wind influences high-rise ventilation operations. *(173)*
12. Discuss the pros and cons of the following high-rise ventilation practices:
    - ventilating below the fire *(175)*
    - ventilating the fire floor *(175)*
    - ventilating above the fire *(176)*
13. Explain the design change that makes modern HVAC systems capable of serving as smoke-control systems. *(177)*
14. List reasons for using an HVAC system to aid in building ventilation. *(177)*
15. List guidelines that firefighters should observe when using an HVAC system to control smoke movement. *(177, 178)*
16. Explain how HVAC systems are shut down, both automatically and manually. *(178)*
17. Discuss the ventilation pros and cons of each of the following built-in ventilation devices:
    - automatic roof vents *(178)*
    - atrium vents *(179)*
    - monitors *(179)*
    - skylights *(179)*
    - curtain boards *(180)*
18. Describe the safest and quickest way to ventilate underground structures such as basements. *(180, 181)*
19. Describe problems associated with ventilating windowless buildings. *(181)*

Questions And Notes

Index

NOTES

NOTES

NOTES

NOTES

NOTES

NOTES

COMMENT SHEET

DATE ____________________ NAME ____

ADDRESS ____

ORGANIZATION REPRESENTED ____

CHAPTER TITLE ____ NUMBER _________

SECTION/PARAGRAPH/FIGURE ______________________________ PAGE ____________

1. Proposal (include proposed wording or identification of wording to be deleted), OR PROPOSED FIGURE:

2. Statement of Problem and Substantiation for Proposal:

RETURN TO: IFSTA Editor
Fire Protection Publications
Oklahoma State University
Stillwater, OK 74078

SIGNATURE ____________________________

Use this sheet to make any suggestions, recommendations, or comments. We need your input to make the manuals as up to date as possible. Your help is appreciated. Use additional pages if necessary.